THE RAT.

THE RAT:

ITS HISTORY & DESTRUCTIVE CHARACTER.

With Numerous Anecdotes.

BY

JAMES RODWELL,

(UNCLE JAMES).

A New Edition.

LONDON:
ROUTLEDGE, WARNE, & ROUTLEDGE,
FARRINGDON STREET;
NEW YORK: 56, WALKER STREET.
1863.

CONTENTS.

PART I.

CHAPTER I.

HE DIFFERENT KINDS OF RATS, AND THEIR NATURAL HISTORY.

CHAPTER II.

THE UNREASONABLE FEAR OF RATS.

CHAPTER III.

GENERAL CHARACTERISTICS OF RATS.

CHAPTER IV.

TAME RATS.

CHAPTER IX.

THIEVISH PROPENSITIES OF RATS.

CHAPTER X.

THE DESTRUCTION AND EXTIRPATION OF RATS.

CHAPTER XI.

WONDERFUL TALES OF RATS.

CHAPTER XII.

COURAGE, FEROCITY, AND CUNNING OF RATS.

CHAPTER XIII.

UNITED ATTACKS OF RATS.

CHAPTER XIV.

THEIR NATURAL WEAPONS.

CHAPTER XV.

ARTICLES MANUFACTURED FROM RAT-SKINS.

CHAPTER XVI.

RATS AS HUMAN FOOD.

CHAPTER XVII.

CHAPTER XVIII.

WHISTLING JOE, THE HERTFORDSHIRE SERMULOT-HUNTER AND RATCATCHER.

CHAPTER XIX.

MISCELLANEOUS ANECDOTES OF SERMULOTS AND SNAILS.

PART II.

TO THE FARMERS OF GREAT BRITAIN, ON THE FECUNDITY AND DEVASTATING CHARACTER OF THE RAT.

CHAPTER I.

ON THE UNIVERSAL PREVALENCE AND DESTRUCTIVE HABITS OF THE RAT.

CHAPTER II.

FECUNDITY OF RATS.

CHAPTER III.

DEVASTATING POWERS OF RATS, AND THEIR ENORMOUS CONSUMPTION OF GRAIN.

CHAPTER IV.

VERMIN-KILLERS AND RAT-MATCHES.

CHAPTER V.

THE SEWER RATS AND RATCATCHERS OF LONDON.

CHAPTER VI.

CHAPTER VII.

RATCATCHERS, AND THE BEST MEANS FOR THE LOCAL EXTIRPATION OF THE RAT.

CHAPTER VIII.

CHAPTER IX.

HOW FARMERS AND OTHERS SHOULD EXTIRPATE THEIR VERMIN.

CHAPTER X.

THE POLICE OF NATURE.

CHAPTER XI.

TRAPPING OF RATS, AND THE VARIOUS KINDS OF RAT-TRAPS.

CHAPTER XII.

POISONING OF RATS.

CHAPTER XIII.

PHOSPHORIC POISONS.

CHAPTER XIV.

MISCELLANEOUS ANECDOTES OF RATS.

PART III.

SOME PROFITABLE HINTS ON THE BREEDING, FEEDING, AND GENERAL MANAGEMENT OF POULTRY.

THE RAT,

AND ITS DEVASTATING CHARACTER.

INTRODUCTION.

For years I have been studying the nature, fecundity, and devastating character of the Rat, and I have spared neither time nor trouble in obtaining all the information within my power. In my opinion it is a subject which materially concerns the general welfare of the human family. Nay, more ; it seems to me extraordinary, that when the cleverest legislators, from time to time, have been exerting their minds and faculties to relieve agricultural and manufacturing depression, the subject of which I am now treating should have remained altogether unnoticed ; for I believe these destructive vermin will be found to be a most fertile source of individual and national distress. In saying this much, it may perhaps be thought that I am attaching too much importance to the subject ; but when the reader shall have perused this work, and given it that deep consideration which is necessary, I believe he will come to the same conclusion as myself.

I must here mention that none of the naturalists whom I have consulted give a clear definition of the rat. Some of them, it is true, present lucid descriptions of it as it is, even to the colour of a hair, or the length and diameter of its intestines ; but as to its general habits, and what it does, they are most vague and imperfect.

Rats feed and labour in the dark ; they shun the approach of man. If we enter a barn or granary, where hundreds are living, we shall not observe any, unless we disturb them in their hiding-places. If we go to a rick that may be one living mass within (a thing by no means uncommon), we shall not see one; or if we dive into a cellar that may be perfectly infested, the result is the same; unless, perchance, a stray one may scud across for a more safe retreat. Hence it is that men seldom think of rats, because they rarely see them; but if rats could by any means be made to live on the surface of the earth instead of in holes and corners, and feed and run about the streets and fields in the open day, like dogs and sheep, the whole nation would be horror-stricken; and ultimately there would not be a man, woman, or child but would have a dog, stick, or gun to effect their destruction, wherever they met with them. Are we to suppose then, because they carry on their ravages in the dark, that they are less destructive? Certainly not. My object then, in making this appeal to my fellow-countrymen, and supplying them with the following History of the Rat (as deduced from the most experienced individuals and naturalists, and interspersed with a host of published and unpublished facts), is for the purpose of rousing them up to one universal warfare against these enemies of mankind, which by their voracious habits contribute to the impoverishing of our farmers and the privation of our fellow-creatures.

In doing this, however, let me be distinctly understood to have in view those rats only that are living and feeding in barns, granaries, ricks, mills, cornfields, warehouses, &c.; in a word, wherever human sustenance is deposited, either on sea or land. Such are their omnivorous, gregarious, and migratory habits, that the whole of Europe, and every temperate region of the known world, are equally infested by them; for when their food runs short, they migrate by hundreds in a drove to some neighbouring barn, granary, rick, or other place, where food is to be found. There will they resort, and scarce anything is proof against them.

Here I may observe that neither bricks, lead, zinc, nor the corners or edges of stone are proof against their teeth and claws. If you examine the holes that they have drilled in your walls, &c., you will find, in most cases, that portions of the bricks or stone on every side are gone,—the space the mortar occupied not being sufficient to admit their bodies. Mr. Waterton tells us, that when the partitionings of his premises were removed, for the purpose of annihilating the rats, he found that where the corners of bricks protruded, so as to obstruct them in their runs, they had actually gnawed them off, and thus obtained a free passage. I have also met with several instances wherein they have not only eaten through zinc drain-pipes, but leaden beer and water pipes. Neither are even gas-pipes free from their depredations. The following instance was lately recorded in the "Manchester Guardian":—

"A circumstance has been communicated to us this week by Messrs. J. P. and E. Westhead and Co. of this town (Manchester), which shows the serious risks of explosion, as well as great annoyance from the smell of escaping gas, may arise out of the ravages of rats. In the new warehouses of that firm in Piccadilly, some of the gas-pipes are placed, as usual, between the floors and ceilings beneath, and an unpleasant smell of gas having been perceived, an examination was instituted into the condition of the pipes, when it was found that some of them had been gnawed through by rats, and in two of them there were holes large enough to permit an escape of gas quite sufficient to cause a most serious and destructive explosion. In one case the hole in the side of the pipe was equal in size to the width of the pipe itself."

The writer then states that these facts should be generally known, as it may be necessary, in some cases, either to have pipes of some harder material than lead, or to case them in tin or sheet iron. But to me it seems far more desirable to get rid of the rats themselves; since there are a host of other things besides gas-pipes that suffer from their ravages. "Yes," it will be said; "but how is this to be done effectually?" To this, I reply, read my book, and then you will ascertain the means.

I can perfectly understand the surprise, nay, I may say, astonishment, created in the minds of some at the weight I attach to rats; but this in most cases arises from want of a due consideration of the subject. It is certain that rats are known to almost every one; but the local injuries they commit are seen only by minute observers. Hence it is, that if a few instances of their swarming numbers and midnight ravages be collected together, and laid open to the understanding, we shall then see one of the greatest causes of agricultural distress.

The old adage says, "that a constant dropping wears a hole in a stone." So is it with rats; for with their astonishing fecundity and omnivorous habits, they are eating like a cancer into the very foundation of British prosperity. Still they are allowed to live, and to revel without any effectual means being resorted to for checking their devastations. But let our farm vermin be destroyed, our hedges levelled, and our lands fully cultivated; and I believe there is little doubt but that our home-grown produce would be more than amply sufficient to supply all our domestic wants. Then, and not till then, do I believe that Great Britain and Ireland will feel themselves to be—what in truth they ought to be—a great and independent people.

PART I.

CHAPTER I.

THE DIFFERENT KINDS OF RATS, AND THEIR NATURAL HISTORY.

I SHALL commence by explaining the natural characteristics of the different kinds of rats; as the WATER VOLE (or Water Rat); the BLACK RAT; the ALBINO (or White Rat); and the BROWN RAT.

*The Water Vole.**

This species of rat bears little resemblance to those just enumerated, either in habits or manners. It is as innocent and seclusive as the others are daring and rapacious. It eats neither fish, flesh, nor fowl, but lives entirely on roots that grow in the water, and at the water's edge; which position it never quits under any circumstances, either in summer or winter; but there it lives and dies, in quiet seclusion. It is perfectly herbivorous, and may often be seen sitting on a stone in the stream, or among the rushes on the bank-side, with a piece of succulent root beween its fore-paws, and nibbling its repast in perfect peace with every living thing. In its external appearance it is not unlike a diminutive otter, but is as timid and innocent in its expression as the otter is fierce and vindictive. It is of a reddish-brown colour, and about the size of a common sewer rat, but with a much shorter tail, which is covered with hair. It seems to have no ears, as they are so short that the fur entirely

* The common Water Vole is the *Arvicola* of Cuvier, who divides the genius into four species: the Water Vole (*Mus amphibius*, Linnæus); the Alsacian Vole (*Mus terrestris*); the Meadow Vole (*Mus arvalis*); and the Economic Vole (*Mus œconomicus*).

hides them; and its front teeth, or rather incisor teeth, are much longer and stronger than those of the other species of rats, and are perfectly yellow. It has young but twice a year, and very few at a birth; consequently the species is rather scarce. The great majority of those animals which we see about the ditches and rivers are the common brown rat.

The Black Rat (*Mus Rattus*, Linnæus).

This species is one-third less than the brown rat, and with ears and tail longer in proportion. Its colour is greyish black. The head is elongated; the muzzle taper and divided, and garnished with numerous long black hairs. The upper jaw projects far beyond the lower, which is remarkably short; the tongue is smooth; the nostrils open and crescent-shaped; the eyes small, but black, and very prominent. It has three grinders in each jaw (the first of which is the largest), and four incisor or front teeth. In its disposition it is fierce, daring, and omnivorous.

Mr. Bell tells us that, independently of devouring every digestible substance, it will not refuse even old shoes or woollen clothes.

Blumenbach informs us that the black rat is extremely voracious; that it eats even scorpions, and follows man and his provisions everywhere, even on board his ships or into the deepest mines. Mr. Bell says that, from its inferiority in size, it has nearly disappeared; that the brown rat has superseded it to such a degree that with us it has become almost extinct.

This circumstance, it appears, has been a source of great pain and grief to our amusing traveller and naturalist Mr. Waterton, who tells us that he once rode fifty miles to see one, and that when he beheld it he could not help exclaiming, "Poor injured Briton! Hard, indeed, has been the fate of thy family! In another generation, at furthest, it will probably sink down to the dust for ever!"

I fear that our worthy friend and patriot has been labouring under two mistakes. In the first place I shall quote Baron Cuvier, who tells us that the black rat, like the *surmulot* or brown rat, appears not to be aboriginal in Europe. No evidence exists of this animal being found among the ancients; and the modern authors, who have spoken clearly upon the subject, go no further back than the sixteenth century.

Gesner is, perhaps, the first naturalist who has described the black rat. Some think, with Linnæus and Pallas, that we received it from America; and others believe that it was a present of our own to that country, made after we had ourselves received it from the Eastern regions.

It is certain that the black rat is to be found in all the warm and temperate climates of the globe; that it is wonderfully common in Persia, and multiplied to a prodigious extent in the Western Islands, where it is not obliged by winter to seek refuge in the habitations of man; but where the fields, through the entire year, present it with abundance of nutriment. In all parts of America, from its ravages and devastations, it has become a perfect scourge. In fact, the black rat consumes an immense quantity of provisions, and destroys and damages still more than it consumes, particularly in the fields, where it tears up plants from the roots, of which it eats but a portion.

Gesner furthermore states that these animals bring forth many times in the year, and that during the season of their amours they have very violent combats, and utter cries resembling a sharp hissing. Their young are born entirely naked, and with their eyes shut. They are generally about nine, or sometimes more, in number.

Sir W. Jardine, in his "Naturalist's Library," believes that this animal was originally imported from the Continent, where it first made its appearance at the beginning of the sixteenth century, and was supposed to come from the East. If so, it may as well be called a Turkish or an Irish rat as an English. Mr. Bell says it is possible, from the proximity of the two countries, that the black rat was first imported into this country from France. Indeed the Welsh name for it, which signifies "French mouse," appears to favour this opinion. The French, on the other hand, call it the English rat, though it is far more numerous in Paris than London.

Most historians, however, agree as to its being a foreigner, whencesoever it came; and there are no records of its existence in this country prior to the sixteenth century. The immense number of 250,000 rats were killed in a few days in the sewers of Paris; and, with the exception of 500 or 600 black rats, the whole of them were of the brown Norwegian species. The black rats do not then live in deadly strife with their brown neighbours; but, on the contrary, fraternize and live on terms of domestic habitude, whence springs a numerous and motley brood of young ones.

Now, let us look to this side of the Channel, and see how far black rats flourish with us; then we can form a tolerably correct notion as to which climate is most genial to their nature, and come to a pretty fair conclusion as to which country the black rats would call their father-land, were they gifted with the powers of articulation. At all events, it is clear that they are no great rarity in Paris. Now, for my own part, I have seen thousands of rats in London, independently of those I have seen in the country; yet I can safely aver that, with the exception of those seen in the "happy family," I never once saw a black rat among them; that is to say, a large-eared, sharp-nosed, fierce-eyed, scaly-tailed, sable-coated *Mus Rattus*. I also add Mr. Waterton's own testimony to the fact of their scarcity in this country; since it cost him so much time, trouble, and expense to gratify his curiosity in seeing one. Nor is he alone in this respect; for the good people of Bristol, some few years ago, were perfectly astonished upon beholding one which had been caught, and sent it up to the Philosophical Institution, where upon examination they pronounced it to be one of the black English rats—"a race which is now nearly extinct, having been all but exterminated by the rats now ordinarily met with."

This, I suppose, was the decision of the philosophers of the institution, which perfectly coincides with Mr. Waterton's views upon the subject. Yet, for my part, I am quite at a loss to know, if England be the natal land of black rats, why they should be more numerous in France than in their native country? or why the brown rat in England should be so barbarous towards the poor native nigger, when, at the same time, in Paris, he places him on a footing of perfect equality.

The opinion that the brown rat is a deadly enemy to the black rat is refuted by the following:—

"Since Louis Philippe left the Tuileries the place has been uninhabited; for a vast multitude of black and brown rats have established an immense colony in the cellars of the once royal castle. Some old shoes, old hats, and some sacks of potatoes, which had been left there, have, up to the present time (1851), amply served them for provisions; and as there is a direct communication between the cellars and the river Seine, they had everything they required to lead a very joyous life. Recently, however, they have been making excursions into the houses of the Rue de Rivoli, and the inhabitants having made a complaint to the Prefect of the Seine, orders were given to the person charged with the destruction of the vermin to organize a razzia against those intruders. It is said that, on entering the cellars, he found a complete mass of these black and brown rats, which formerly were said to be mortal enemies, but now are living on fraternal terms; and, in consequence of crossing the breeds, many of them were dark on the backs, with white bellies and tails. The skins of this race are considered valuable. The night before last the ratcatcher of the capital commenced setting his traps, and on the following morning he had caught 847 rats. According to custom, their tails were cut off, and sent to the Hôtel de Ville, in order to support the claim for the usual reward."

This, I think, is tolerably clear evidence. And now let us return to our worthy friend Mr. Waterton, and the gentlemen of the Philosophical Institution, who say that the brown rat has worried nearly every individual of the original black rat of Great Britain. This, I fear, is Mr. Waterton's second mistake, since their diminution may be attributed to more natural causes. The black rat is one-third less than the brown rat, consequently much weaker. The result is, that in the breeding seasons the stronger male rats beat off the weaker, and run away with the black ladies, who in due time have a family of half-bred young ones. When these have become matured, they breed in again with the brown rats; their young become much lighter, and so on for a few generations, and the entire breed will become confirmed brown rats; while the old black rats, having lived as long as nature will allow them, cease to exist, and the race becomes extinct.

Albinos, or White Rats.

Before I give an account of the Brown rat, I must notice the Albino, or White rat. Some authors believe it is a mere accidental variation of the brown rat. Mr. Richardson, in his "Pests of the Farm," also expresses the same opinion. He says, "The common brown rat sometimes presents albino characteristics; that is to say, it is occasionally to be found of a white colour, with red eyes. In the neighbourhood of Greenock, for instance, he says, there were, some years ago, numbers of these albinos to be met with, especially among the shipping. Some specimens were sent him, and he kept one of them as a pet for a considerable time."

A colony of white rats was lately discovered in the Ainsworth Colliery, near Bury. They committed great depredations, when they had an opportunity, upon the food of the colliers.

In London, at the present time, these animals, being bred for fancy, are becoming very numerous, and sell at the rate of four shillings a couple. In shape and manners they are exactly like the common barn rat, but rather smaller. Their fur in every part is purest white; and their eyes, noses, and skin, beneath the fur, are of a most delicate pink.

The first who bred these for fancy and profit was a person by the name of Ostin, residing in the Waterloo Road; and he was the first man who brought to perfection the happy family, which may be seen daily at the foot of Waterloo Bridge, London. He informed me that he first procured two white rats, male and female, from Normandy about three years ago; and from this couple he has bred an immense number in cages. He has also initiated his son-in-law in the art of subduing the natural cravings of various animals, and reducing them to one standard of peace and equality. He exhibits his happy family every fine evening in Regent Street, and in the present instance is my principal informant. At the time I write he has above a hundred white rats, besides others. He has crossed the breed with the brown and black rat; and has produced a vast number of both brown and white, and black and white piebald young ones, which are pretty little creatures, and as tame as kittens. He says, "they breed six times in the year; and when the young are two weeks old, the mother is again pregnant. The young ones will breed at four months old." He mentioned one female which bred so fast that she died from sheer exhaustion. But the natural powers of the rat for breeding are so great, that I believe few animals, if any, in the creation can equal them. He also informs me that these animals are subject to no diseases, except when kept in a dirty cage for any length of time; and then, like ferrets, they are subject to a kind of mange; but cleanliness and good diet will soon cure them. The most he has had in a litter were thirteen, and not a dark hair among them.

Now this does not favour the general opinion of naturalists, or of Mr. Richardson, as to the albino or white rat being an accidental variation from the brown or black rat, because it is a well-known fact to all breeders, either of beasts or birds, that any young one, which may accidentally vary in colour from the rest, will in breeding-time revert to the original stock. So far from doing this, the albino, or white rat, will breed for generations together without varying in colour. Consequently I am led to believe that they are a *bonâ fide* species; but this I will most humbly leave in the hands of profound naturalists to investigate and decide.

The Brown Rat (*Mus decumanus*, Linnæus).

We now come to the common Brown rat, or, as Buffon and Cuvier style it, the *surmulot.* To me it is of little import by what name they call it. This is the animal against whose ravages the present work is directed. Its natural characteristics are so well known that a description seems almost superfluous. Nevertheless, for the greater completion of my work, I will give its colour, parts, and proportions, as described by Linnæus.

The brown rat is the largest species of the genus that occurs with us. Its body is rather elongated and full, the limbs short and moderately strong, the neck short, the head of moderate size, compressed, and rather pointed; the ears are short and round, the tail long, tapering to a point, and covered with 200 rows of scales. On the fore feet are four toes, of which the two middle are much the longest; the soles are bare, and have five prominent papillæ. The hind feet have five toes, of which the three middle are the longest, and nearly equal, the first shorter than the fifth; the sole is bare up to the heel, and has six papillæ. The general colour of the upper parts is reddish brown; the long hairs are black at the end, the lower parts greyish white. On the feet the hairs are very short, whitish, and glistening; the claws are horn-coloured, or greyish yellow. The œsophagus is four inches long; the stomach transversely oblong, $2\frac{1}{2}$ inches in length; the intestine slender, about $2\frac{1}{2}$ twelfths in diameter for four feet three inches; it then enters a large curved sack formed by the head of the colon, which projects two inches, with a diameter of nine-twelfths; from thence to the extremity the intestine measures ten inches; its diameter at first seven-twelfths, but gradually diminishing to four-twelfths. The liver is divided into six lobes, and there is no gall-bladder. In the female there are six pectoral and six inguino-ventral mammæ.

	Male.		Female.
Length to the end of tail	19 inches	..	16 inches.
Head	$2\frac{3}{4}$,,	..	$2\frac{1}{4}$,,
Tail	$8\frac{1}{4}$,,	..	$7\frac{1}{4}$,,
Ears	$\frac{3}{4}$,,	..	$\frac{3}{4}$,,
Hind feet	$1\frac{3}{4}$,,	..	$1\frac{1}{2}$,,

Besides the black and brown, Baron Cuvier gives accounts of seventy-two different kinds of rats, each of which has its native locality, and which it seldom or never quits, except by force or accident. But the black and brown rats are citizens of every genial portion of the globe, and seem to say the world is theirs, for they go where they like, and do as they please. Now it may be asked from whence came they? Ay, there's the rub; for I know of no animals, in the whole range of natural history, wherein there is so much discrepancy of opinion as to the land of their nativity, or such conflicting testimony adduced by the various philosophers as to which country has the honour of claiming these little truants as its legitimate offspring.* Some

* Cuvier says that this animal did not pass into Europe till the

naturalists believe these came from the East Indies; others believe they came from the West. Many assert they came from Norway, while others maintain that they were common in England before the Norwegians even heard of them.

It may surprise those who are sticklers for the Scandinavian origin, to know that this rat was brought to England from the Indies and Persia in 1730; that in 1750 the breed made its way to France, and its progress over Europe has since then been more or less rapid; and that when Pallas was travelling in Southern Russia he saw the first detachment arrive, near the mouth of the Volga, in 1766.

Some respectable authorities state that the brown rat came from Persia and the southern regions of Asia, and that the fact is rendered sufficiently evident from the testimonies of Pallas and F. Cuvier. Pallas describes the migratory nature of these rats, and states that in the autumn of 1729 they arrived at Astrachan, in Russia, in such incredible numbers that nothing could be done to oppose them. They came from the western deserts, and even the waves of the Volga did not arrest their progress.

It is said by others that their first arrival was on the coast of Ireland, in those ships that used to trade in provisions to Gibraltar, and that perhaps we owe to a single couple of these animals the numerous progeny now infesting the whole extent of the British empire. Mr. Newman asserts that we received the rat from Hanover, whence it was called the Hanoverian rat. Mr. Waterton states that his father, who was a naturalist, always maintained that they came to us in the very ship which brought George I. to England, and that they were seen swimming in a shoal from the ship to the shore. Pennant says that the brown rat arrived in England about 1728, and in Paris twenty years later; but a modern writer asserts that they appeared in France in the middle of the sixteenth century, and were first observed in Paris. Buffon says that it is uncertain from whence they came, though it was only ten years before, that they arrived in France, and this I believe to be about the true state of the case; though the Egyptians maintain that they

eighteenth century. He further observes that it appears to belong to Persia, where it lives in burrows, and that it was not till 1727 that, after an earthquake, it arrived at Astrachan, by crossing the Volga.

were made out of the mud of the Nile, and assert that they have seen them in the process of formation, being half rat, half mud.

After all, it matters little from whence rats came. Here they are ; and how to get rid of them will form the subject of the following pages.

CHAPTER II.

THE UNREASONABLE FEAR OF RATS.

I KNOW of no one thing so universally detested, or so unjustly charged with everything that is foul, treacherous, and disgusting, as the rat. I say unjustly, because whatever it does, like every other animal, it is only following the bent of its nature. But, at the same time, I believe a vast amount of the disgust exhibited at the bare mention of its name to be mere affectation. The most striking instance I ever met with took place one evening in London, where a friend of mine supped, or rather was to have supped, with a party by invitation, the good lady having invited her friends, in the temporary absence of her husband. I here give his own account.

The party met, and as he was the only stranger present, of course the formality of introducing him to each was indispensable. This ceremony being concluded, the supper was served up. There were roast ducks, fowls roast and boiled, plovers, curried rabbits, ham, lamb, &c., with vegetables of all kinds, and soups and sauces in profusion. They were all seated round the table in pairs, to please the peculiar notions of the amiable hostess. There were four Miss So-and-So's, of different surnames, and four Master So-and-So's to correspond. They were packed up two and two, and there was plenty of everything but room. Then there were Mr. and Mrs. Tweedle, and Mr. and Mrs. Button. These four, as a matter of course, claimed their matrimonial prerogatives. The good landlady herself was seated at the head of the table ; thus making a comfortable pic-nic party of fifteen. The covers were removed, and the steam rose in one vast volume, to the evident satisfaction of all around. The fowls were disjointed, and they were all busily engaged, when there was one of the most tremendous squeals below stairs ever heard. All operations instantly ceased, and every one stared with breathless astonishment. Presently the hostess and my friend went to the door to ascertain the cause, when they found a terrible rushing up stairs. It was her hopeful son and heir, Master Bobby. "O mother," said he, "see what a jolly great rat I've caught!" at the same time holding up an enormous fellow fast gripped in a gin-trap. She slammed the door, and uttering a faint shriek, staggered backwards into her chair, and fainted away, or at least seemed to faint. All the party were on their legs, the ladies clinging fast to the gentlemen for protection. My friend of course tried all the little antidotes of which he was master; such as bathing the lady's hands and temples with vinegar, pump-water, eau-de-Cologne, &c.; but all to no purpose. Yet, what made the matter still more appalling was, that Miss So-and-So was fainting in the arms of Mr. So-and-So; then her companion followed suit, and so the ladies dropped off one after the other, till it was quite clear that fainting was the order of the evening. As for poor Mr. Button, though himself a very diminutive gentleman, he was nevertheless the sole proprietor of a very large wife and six little Buttons, and in his ferocity a perfect lion-eater. His gaunt angel had fainted away at full length on the sofa, which roused him to the highest pitch of daring; for, in his vengeance, he seized a mahogany chair to smash the enemy, and in brandishing it above his head one of the legs came in contact with a splendid mirror, and made a brilliant star in the centre; but, unmindful of the accident, he still proceeded; and whether in imagination he saw the rat, or whether to frighten it I know not, but down he brought the chair with such a tremendous crash that off flew the legs, and there it sat bolt upright, like an old Greenwich pensioner without his stumps. My friend's inward laughter, however, was soon turned to something more serious, for his fat damsel, as if to show them all how to faint, raised her hands, and giving a most awful scream, threw herself plump into his arms, which capsized him, table, supper, and all. He fell across the upper legs, which flew off, and there he lay puffing and blowing on the broad of his back, with this precious grampus on top of him. All was darkness, when up came the big servant, followed by a lesser one with a candle; in she went blundering, and fell flat on top of them. That was too much; but clumsy as she was, she soon scrambled up, and released him from his counterpane by rolling her over, at which, he says, she gave a groan that felt for all the world like the threatening rumbles of an angry volcano. Our hero, however, had no sooner gained his legs, than there was a rat-a-tat-tat-a-tat-tat at the door, followed by a ring-a-ding-ding-a-ding-ding. "Lorks, marm," said the big servant, "if there ain't master!—who'd have thought it?" It has often been said that what is one person's meat is another's poison, so it turned out in this case; for the servant's announcement, which made her mistress faint in reality, had the effect of bringing every one else to their senses; and even the fat damsel rolled and kicked till she got on to her hands and knees, when my friend, by a powerful effort of physical exertion, succeeded in performing a chivalrous act by helping her on her feet.

The master, who was altogether unexpected, was a man of asthmatical and acetic disposition, and had a somewhat quaint and dramatic way of expressing himself; yet, unlike his amiable wife, he had a great aversion to company, since he looked upon it as the annihilator of domestic felicity. He no sooner entered and saw the real state of affairs, than, with all the brevity of Lady Macbeth, he requested the company to stand not upon the order of going, but to go at once! To which request, with all the etiquette of good breeding, my friend responded by taking his hat, and making his bow, never more to return.

Now, here was a scene of pitiable disasters, arising out of a little nonsensical display at seeing a dead rat! Had the mistress given Master Bobby a cuff on the ear, and ordered him to throw it away, she would not only have shown herself a wiser woman, but have saved all the misery and expense which otherwise resulted.

The fear of rats exhibits a mere childish weakness, which in parents, who ought to teach their children better, is highly reprehensible. What harm do they think a rat will do them, which is already more frightened than they are, and only wants to effect its escape. It is true that in some instances, if you press a rat into a corner, and will not let it escape, it will fly at you; yet it will not do so, if you open the door and let it out, or let it gain its hole. But if you pursue it to desperation, then, like almost every other animal, it will fight for its life; and you, like every other sportsman, must put up with the consequences of the chase. Here we may apply the old adage, which says that if we tread upon a worm, it will turn again.

A few years back I attended a rat-match in London, at which the dog which could kill the greatest number in the shortest time would win the prize. The first man that entered the pit brought in with him a dog, which was as handsome as the man himself was ugly. Time being called, he seized his dog by each side of the face, and, arching his long carcase, was soon in readiness. They now made a curious picture. From the fierce anxiety of their countenances, it became a strong matter of doubt as to which would mouth the first rat—the dog or its master. However, upon the signal being given, away went the dog, first killing one rat, then another; down went the man on his hands and knees, then pounded the floor most furiously, and roared and bellowed with all his might, to urge on the dog. The rats were falling in every direction, when, all in an instant, the man stood bolt upright, with his eyes staring like a madman, and his mouth wide open. But the cause, to the great amusement of all present, soon became apparent. It is the custom for those who enter the pit to tie a piece of string or garter round each ankle, to prevent the rats from crawling up their legs beneath their trousers. He had neglected to do this, and a rat was plainly seen working its way up his body between his skin and his shirt. With maniacal desperation he pulled off his cravat, and, tearing open his shirt, exposed his thin scraggy neck. Presently out came a pretty little glossy creature on his shoulder, and made a spring to the edge of the pit, which it succeeded in accomplishing. Heels-over-head went a dozen or twenty of the lookers-on, forms and all; and from the general scrambling, kicking, bustling, and alarm, one might have thought that, had it not been for their hats and boots, the poor little frightened rat would have swallowed at least a dozen of them; but, as it was, the little creature made its escape; and thus were they allowed to return home to their families undigested. Suffice it to say, I have since heard that nothing could ever induce this man to enter the pit again, but that he always pays another to do it for him.

On the opposite side to where the rat made its escape sat an enormous publican, who had laughed most heartily at the discomfiture of his friends. His face was a perfect picture of the sign of his house, namely, the "Rising Sun." He was lounging carelessly on the edge of the pit, and resting his chin on his thumbs, when in came the second dog, —a fine furious creature. Time was called; the dog set to work, and down lay the rats, one after the other, with a single bite each. Presently one seized the dog by the lip; he gave his head a violent shake, and twirled the rat into the fat publican's face. To describe his agitation and alarm would be impossible; but, throwing his head back, the rat fell into the bosom of his coat; and, in his anxiety to get it unbuttoned, he puffed, grunted, and blew like a great hog with a bone in its throat; and thus ended his sport for that evening.

Unhappily, however, these rat-frights do not always terminate so harmlessly as in the preceding cases. A friend of mine once informed me that twenty years since his father took a house in Edinburgh, and that after he had taken it, he found, to his dismay, that it was swarming with rats. However, one day, as they were all in the kitchen, where the boards of the flooring were about an inch apart, they were suddenly aroused by two rats, which had commenced a regular battle beneath the boards. My informant told me that his little brother became very much alarmed; when suddenly one of the rats gave a dreadful squeal, and at the same time one of their hind legs and a tail appeared through the cracks, which so frightened the lad, that he sprang to the other end of the place, when it was found, to the great affliction of the family, that he was bereft of reason, or, rather, he had become a complete maniac; nor was it till some weeks had passed, accompanied with sound medical treatment and warm baths, that anything like consciousness returned. However, by degrees he recovered the possession of his faculties; but to this day he is horror-stricken at the bare mention of a rat.

This is only one case out of many thousands that might be mentioned, where consequences infinitely more serious than even this have frequently occurred through this foolish fear of rats.

The "Presse," of Paris, some time ago related an extraordinary case of death from fright. A young woman was passing near the Rue Cadet, when she suddenly fell to the ground, exclaiming "The rat! the rat!" At first nobody could comprehend the meaning of her exclamations; but on being taken into a druggist's shop, and placed on a chair, a rat was seen to run from beneath her gown. It was then evident that the rat, which had come from a sewer just as she was passing, had got between her legs, and that, when she fell from fright, it had concealed itself under her clothing. She was taken home to her friends, in a state of delirium, which lasted four days, during which time the only words she uttered were "The rat! the rat!" but on the evening of the fourth day she expired.

Now here was a melancholy occurrence arising out of this immoderate fear of rats. What had the rat done to her? Nothing whatever, except hiding in her clothes, and making its escape as soon as possible. Yet from the veriest fear she becomes deranged, and dies a maniac. I would that every female, and man too, were as brave and resolute as Mary Ann Gymer, who, at the police-office, stated that on the previous morning, while returning from market, the prisoner (whom she had given into custody) came behind her, and placed a live rat on her right shoulder. On turning her head, the creature made a bite at her face; but the instant he let go its tail it sprang to the ground, and made its escape down an area grating. The prisoner stood laughing heartily at the outrage; but such was her indignation at the daring of the fellow, that she ran up to him, and dealt him a most violent blow in the face with her pattens. The magistrate said he was not at all surprised at her inflicting summary vengeance on a man who had behaved so disgracefully towards her. The placing such a thing as a live rat upon her shoulder was enough to alarm her, and upon a more nervous person might have been attended with the most melancholy consequences. The magistrate gave the prisoner a severe admonition as to his future conduct, and assured him that, had it not been for her intercession in his behalf, he most assuredly should have fixed a very heavy fine upon him.

CHAPTER III.

GENERAL CHARACTERISTICS OF RATS,—THEIR DISEASES, FEROCIOUS CANNIBALISM, NATURAL AFFECTION, &c.

I CANNOT ascertain any disease rats are subject to, except the one we have so much cause to complain of, namely, consumption of food; and for all I have seen, they have, without an exception, been fine fat fellows, with nothing in the shape of disease about them; yet, if kept in close confinement and dirty, then, like every other animal under similar circumstances, they are subject to a kind of surfeit or mange; but in a state of freedom I believe they are subject to no natural malady. Nevertheless they are the finest and most effectual physicians in the world; for should any of their fraternity be mopish and dull from pain or sickness, arising out of accident, old age, or what not, they cure all their maladies at once by eating them up. At the same time they are the most peaceable of all republics, for should any internal quarrels and fights arise, they all gather round the combatants, and no matter who wins or loses, or what the cause of contention, they put an end to the feud by tearing them in pieces, and transporting the pair of them down their throats; and thus is peace most speedily and effectually restored.

This, then, may account for the healthy appearance of rats, since they instantly and so effectually dispose of their sick and ailing; and it may be no stretch of the imagination to infer that few die a natural death; but where such is the case, they will leave all other food to dispose of their dead. Hence has arisen the proverb I have heard, that in barns and ricks dead rats are almost as rare as dead donkeys; though, at the same time, there are instances on record of their filial and parental affection, and attachments as strong as is to be found among animals of any other class, not omitting even the human family; but should we at any time see a dead rat lying about, we may be almost certain that it has been very recently killed by some animal, trap, accident, or poison; but if left where it lies, the morrow's sun will seldom or never shine upon it.

The most determined instance of cannibalism among rats I ever witnessed took place about thirty years ago, at Wareside, where I was at school. The Principal was a real old English schoolmaster, and the very antipodes of your modern pedagogues. He had no idea of storing our heads at the expense of our healths, or sending us home at Midsummer and Christmas a race of upstart, pigmy, juvenile men. His principle was to well fill our stomachs with the solids of life, and then, if any vacancy remained, to eke out with reading, writing, and arithmetic; but the latter only went as far as the rule of three—he cared nothing for anything beyond that, as it formed the boundary of his arithmetical knowledge; and since it had served his every purpose up to a good old age, why, of course, it was sufficient for any one else, and who dared dispute with a British schoolmaster in those days? His terms were tolerably reasonable, and to supply his establishment as economically as possible, he kept a farm, whereon he used to raise nearly everything for the school's consumption, as well as supply the market on Saturdays.

The cows were wont to be brought from the farm to be milked in a cowhouse adjoining the play-ground; and against which, on the opposite side, stood a large faggot-stack. Now, it was Milly the housemaid's business to milk these said cows. Milly was a pretty, laughing, dark-eyed, kind-hearted, curly-headed creature—one for whom nature had done much, but education nothing; yet, nevertheless, her cheerful, single-hearted, innocent hilarity would rivet the affections even of the most serious. Suffice it to say, that with Milly I was an especial favourite; consequently I used to come in for many little attentions and favours; such, for instance, as having a good fill of new milk sucked through a straw when master was out.

One afternoon, when on one of these succulent expeditions, I experienced two awful frights; for scarcely had I secreted myself in the cowhouse when I heard the master's footsteps advancing from the adjoining stables. You may imagine my fright at being caught out of the boundaries, when a retreat was impossible. Nor do I believe that our black cat felt more confounded and astonished when she missed the sparrow, and went souse into the water-butt, than I did at the terribly well-known grunting cough of this grand master of the rod. I thought he had gone to market, and I am satisfied that pussy did not scramble out of the water with greater agility than I clambered into the loft above the calf-pen, and as speedily buried myself among the straw. Nor was it till I heard the shuffling tramp of his heavy step and capacious slippers die gradually away, that I dared to breathe, or think myself something living.

While thus listening and musing, with all the frantic bogies of imagination dancing menacingly before my mental vision, I was suddenly aroused from my unenviable reverie by the grumbling and squealing of rats in the adjoining faggot-stack; when, through a hole in the boards, in tumbled a pair of monsters, rolly-polly over each other, and fighting like two bull-dogs. In came others in all directions, till the place seemed one mass of living rats. They all gathered close round the combatants—those behind scrambling and pushing on those before them, till there was not a ring a foot in diameter left to fight in. Those in front, and those immediately next them, were reared upon their hind quarters. By this time the stronger had got the weaker down, and was in the act of killing him, when his dying moans seemed to be the invitation for a general onslaught. They then, one and all, as if by a given signal, fell upon the combatants, and scrambled over each other's backs—those behind struggling hard to be in at the death, till nothing of the victims was seen. Indeed they looked like a mound of rats, or all backs and tails. Presently there was a most violent and general struggle; so that you would have thought they were all fighting together. When one ran away, he was followed by others scrambling for a piece. Then a second, a third, and so on, till the combatants were torn to tatters; and then came the quarrelling, grumbling, and scranching of bones—'twas enough to make one's hair stand on end. No sooner were they demolished, than in came a large fellow, evidently attracted either by the smell of blood, or sound of fighting, when, meeting a fellow-rat somewhat besmeared in the fray, he, without any ceremony, fell furiously upon him, and there was a second battle. The rest gathered round, the same as before, doubtless in hopes of having a second feast, when in came Milly, whose sudden appearance put them all to flight. She threw down the pails, and away she ran screaming with all her might. Now, thought I, is the time for my escape. Down I jumped, and seized a stick. In came the boys, heels over head, and I was busily engaged banging away at everything within reach, when in came the master, inquiring where the rats were. I told him they had passed through a hole into the faggot-stack. Out they all ran in pursuit; but no one thought to ask me whence I came, and so I got out of my difficulty.

Thus it appears evident that rats do not cluster round these outbreaks for the purposes of peace, but to gratify a carnivorous appetite for hot blood; and that they will greedily indulge this gloating propensity, whenever an opportunity offers, with any of the smaller animals as well as their own species.

There is a notion abroad that rats, when caught in a trap by the foot or leg, will sometimes escape by gnawing away the limb above the trap. This I believe to be perfectly erroneous, and I am supported in my opinion by a gentleman, where, speaking of the rats of Shropshire, he says: "So savage and voracious are the common Norway rats, that often and often, when one of these gentry is caught in a trap, the others attack and eat him up; and frequently the keepers find from ten to twenty rats caught in the rabbit-traps during the night, though set far away from ricks or buildings of any kind; and perhaps two-thirds of them before morning would be eaten by these cannibals of the worst kind; for," he says, "they do not wait even to kill their brother rats in trouble before they feast upon them."

One evening I called upon an acquaintance of mine to obtain some particular information, and found him just going to decide a wager respecting a large male ferret of the polecat breed, which was to destroy fifty rats within the hour. It must be borne in mind that this ferret was trained for the purpose. The rats were placed in a large square space measuring eight or ten feet from corner to corner. The ferret was put in, and it was astonishing to see the systematic way in which he set about his work. Some of the larger rats were very great cowards, and surrendered with scarcely a struggle; while some of the smaller, or three-parts-grown ones, fought most desperately. One of these drew my particular attention. The ferret, in making his attacks, was beaten off several times, to his great discomfiture; for the rat bit him most severely. At last the ferret bustled the fight and succeeded in getting the rat upon its back, with one of his feet upon the lower part of its belly. In this position they remained for some minutes, with their heads close to each other, and their mouths wide open. The ferret was rather exhausted with his former conflicts, and every move he made the rat bit him. At last he lost his temper, and making one desperate effort, he succeeded in getting the rat within his deadly grasp. He threw himself upon his side, and, cuddling the rat close to him, he fixed his teeth in its neck. While thus engaged, a rat was running carelessly about; all at once, when near the ferret, it threw up its head, as if a new idea had struck it; it retreated till it met with another, and it was astonishing to see the instantaneous effect produced in the second. Off they ran together to the corner where the ferret lay. The fact was, they scented the blood of either the rat or ferret, which in both was running in profusion. Without any further ceremony they seized the ferret fast by the crown of the head, and drew themselves up for a comfortable suck of warm blood. The ferret, feeling the smart, thought it was his old opponent that was struggling in his grasp, and bit his lifeless victim most furiously. Presently he let go the dead rat, and seemed astounded at the audacity of the others. He began to struggle, and they seemed quite offended at being disturbed at their repast. He very soon, however, succeeded in catching hold of one of them, and the other ran away, but only for a few seconds. The ferret demolished the whole fifty considerably under the hour. Nevertheless, two facts were established beyond a doubt—first, that rats are perfectly carnivorous; and, secondly, that they delight in sucking hot blood.

Having thus given some accounts of the worst propensities of rats, it is but fair to present some authenticated facts, as well as my own personal observation, in confirmation of their better qualities.

Mr. Bell quotes a case of fraternal affection among rats from Mr. Jesse. A gentleman was walking out in the meadows one evening, and observed a number of rats in the act of migrating from one place to another, which it is well known they are in the habit of doing occasionally. He stood perfectly still while the whole assembly passed close to him. His astonishment, however, was great when he saw an old blind rat, which held a piece of stick at one end in its mouth, while another rat had hold of the other end of it, and thus conducted its blind companion. He also says that it is very evident, from several instances, that the rat is not insensible to kindness, and that it may be powerfully attached to those who feed and caress it.

The Rev. W. Cotton gives an instance of fidelity among rats:—"On a bright moonlight evening, we discovered two rats on the plank coming into the ship. The foremost was leading the other by a straw—one end of which each held in its mouth. We managed to capture them both, and found, to our surprise, that the one led by the other was stone blind. His faithful friend was trying to get him on board, where he would have comfortable quarters during a three years' cruise."

The maternal affection of rats for their young is not, perhaps, to be surpassed by any other animal; and so far from their being the low, degraded, dirty, ignoble creatures that many imagine them to be, they are, on the contrary, perfectly aristocratic in their habits and notions. Sir W. Jardine says: "The rat is a very cleanly animal; for even when its residence is in a ditch, or sewer, in the midst of all sorts of filth, it almost invariably preserves itself from pollution; and in parts remote from towns its fur is often possessed of considerable beauty. Although, on account of the injury it inflicts upon us, and the abhorrence with which in childhood we are taught to regard it, few persons will be apt to discover much beauty in a rat; nevertheless, any one who has taken notice of rats, can bear testimony to the fact, that in all their leisure time they are constantly sitting on end cleaning their fur, and seem perfectly restless and unhappy till their jackets are dry and clean, and arranged in proper order."

In the spring the rats leave their winter establishments, and mostly repair to some watering-place to spend thesummer months. Here the mother teaches her young the recreations of swimming, fishing, and hunting.

But in their more infantine days she is one of the kindest of nurses, eternally washing their little faces, backs, bellies, legs, and feet, by rolling them from side to side, and licking them over with all the tenderness and solicitude of any other mother. But if an enemy intrudes, she will protect them with all the vicious determination of a tigress, and if she does not succeed in beating him off, she will relinquish the contest only with her life. So, if the old rat should call, who in some cases is a barbarous old brute, she will show him her teeth, and squinny at him till he decamps; but, should she be from home, the infanticidal old cannibal will sometimes eat up her children, and then walk doggedly to his retreat, and lay himself down most tranquilly to digest them.

Two men in a boat were gathering rushes on the borders of the Avon, when a water-rat entered the boat, arranged some of the rushes, and gave birth to seven young ones, which the men at once destroyed. The rat immediately set up such piteous cries, that they endeavoured to drive her from the boat, but she would not go; so they killed her also.

Doubtless she sought this asylum to save her young from the jaws of the old tyrant, and so lost her own life, which says but little for her destroyers, where, from first to last, such a confiding appeal was made to their humanity. But let it be borne in mind that the male rat is not the only animal that will devour the young of its own species. Pigs, both male and female, will sometimes do it; as also tom-cats and rabbits. But, in a wild state, the doe rabbit always goes a considerable distance from the main burrows, where the buck is not likely to travel. Here she makes a hole some two feet long, and deposits her young, which, when she leaves to go abroad for food, she always covers the entrance close up with earth, so that the buck is not likely to find them in her absence. As to boar-pigs, it is well known that if they come across a litter of young pigs, and the opportunity offers, they will chump them up like sweetmeats. Consequently the male rat, however disgusting, must not be individually condemned for murder and cannibalism.

A writer in the "Zoologist" says that he was once an eyewitness to an act of affection on the part of a female rat; which he thought worth recording, more especially as the rat is considered to have little in its character to recommend it. Some persons, who were cutting a field of barley, mowed over a rat's nest full of young ones, when the mother, who was suckling them, instead of running away, remained in the nest, and, in her anxiety for their preservation, actually laid so fast hold of the scythe that she was obliged to be shaken off. This nest was made in a slight depression of the ground, and not under the ground, as usually is the case.

The same gentleman gives another instance of considerable cunning and courage evinced by a water-rat. He was walking by a brook one day, and saw a water-rat run past on the opposite bank in great haste. Almost immediately afterwards came a very fine stoat, hot in pursuit, but evidently running by scent. Backwards and forwards ran both animals within a certain space for upwards of ten minutes, when both made a dead pause within a yard of each other, and he expected to see the rat fall a prey every moment. But such was not the case; for, in an instant, she rushed forward upon the stoat with such open-mouthed fury, that he ran away, and she in turn became the pursuer; nor was she content until she had driven her voracious enemy fairly out of the neighbourhood. There is no doubt that the rat was a female which had young, and was prompted by maternal affection to display the courage she did. But what surprised him was, that she never retreated to a hole, or dived under water, which would have been an almost certain mode of escaping danger; but that would not have prevented her wily enemy from scenting out her young.

The same writer gives another instance of intense affection in the rat. Some gintraps were set for the purpose of taking vermin. On the following morning a large female rat was discovered in one of them, caught by one of her fore legs, but squatting over a nest containing six young ones. The poor animal, regardless of all pain, during the previous night had actually, with the fore paw which was at liberty, and probably with the assistance of the hind feet, contrived to scrape together a quantity of the neighbouring grass, and formed the nest,—thus providing for the warmth and comfort of her young, although she was tortured with iron teeth, and almost disabled by her position in the trap.

Of the unqualified affection of a rat for her young, I was witness to a most interesting and curious instance. I had a sort of compound collection, half aviary and half menagerie. My stock was composed of rabbits, pigeons, ferrets, fowls, cats, dogs, white mice, hedgehogs, guinea-pigs, and canaries. Besides these there was a host of native song-birds and a cock pheasant. I did not keep them like the happy family, all together, but in separate departments. Among these I had an enormous polecat ferret, blind in one eye. He was perhaps the largest ferret I ever saw, and was so tame and attached, that he would follow me in the streets, or anywhere else, like a little dog. In the fields I often used to amuse myself by running away, and giving him the trouble to find me out; still I never was afraid of losing him, because he always wore a small collar round his neck, with a little bell attached to it. However, we were out together one summer's evening, in a meadow, which on one end and side is skirted by a river, and on the other by thick hedges. I was lounging carelessly on, when I heard the ferret make an extraordinary loud chattering noise, something between a cackle and a bark, or rather just such a noise as a monkey will make, when some mischievous boys have his tail through the cage, and are tying it tight in a knot—I instantly ran back, and found him in the ditch, in a state of perfect confusion, and bleeding terribly from the nose. I fancied I saw something disappear, but what I could not tell; yet seeing him bleed so profusely, I imagined he must have run a spike into his nose, having recently seen some set in a game-preserve for the purpose of killing dogs as they jumped through the gaps in the hedges. I paused for a moment, and soon found by his action that there was some game at hand. Presently he sniffed about, and made his way carefully to a bundle of dried grass, leaves, &c., in the hedge-bottom at the root of a bush. When, quick as lightning, out dashed a rat at him, and as quickly disappeared. But what with the smart of the bite, and the force she came with, it threw him fairly on his back. Oh, thought I, here's some sport! He soon recovered himself, though bleeding from a second wound. He made another attempt with the same result, and another bleeding wound. Thus he approached six or seven times with the same consequences. Whether or not she got on the blind side of him I cannot say; but with the stick I had in my hand I determined on dislodging her ladyship; so with the hookey end I forthwith turned over the bundle of dried grass, &c., which parted in the middle, and lifted up like the lid of a box; when lo, and behold! there she sat fondling over a host of naked, blind young ones, about three or four days old. The sudden appearance of daylight seemed at first to bewilder the poor creature; but she soon recovered, and began licking her offspring, yet looked unconscious of what she was doing, for her eyes were fixed most piercingly on the ferret. He was sniffing about, yet creeping stealthily nearer and nearer; but when he came within a certain distance, out she dashed at him, and knocked him over again; this she repeated three or four times, each time inflicting a fresh wound, until the ferret was bleeding from all points, which made him extremely cautious; and it was some time before he would venture again within certain limits. In the mean time the poor rat was licking and fondling over her young, as if to persuade them that they were all safe, and that there was no danger at hand. By this time, the ferret, gathering himself up for mischief, pressed boldly forward—when out she dashed at him; but two of her young ones were hanging to her teats, which I suppose broke her spring; they fell off, helplessly sprawling on their backs, and she got into the ferret's clutches. He had seized her by the skin of the back, and was cuddling her up in his deadly grasp, which was too much for me. The idea of so spirited a creature, which had beat off a much larger and deadly enemy so many times, becoming a prey to his fangs, and leaving twelve poor little blind sucklings without a mother, needed no further argument; so into the ditch I jumped, and taking the ferret by the tail, laid them on the bank, and thrust the small end of my stick into his mouth, and, by giving it a slight wrench, allowed the rat to get loose; but, instead of running off directly to her young as I expected, she turned upon the ferret again and again, a perfect little vixen, and bit him most severely before I could get out of the ditch: it was very evident that she had made up her mind to a life-and-death battle. I kept hold of the ferret's tail, which very much impeded his action; but no sooner was I fairly landed than I suspended him in the air at arm's length. The rat sprang up five or six times, but could not reach him; when, all in an instant, like a squirrel, she ran up my leg and body, then along my arm, and dropping on him, gave him another bite and fell. This I thought anything but fair play; so when she attempted it a second time I brushed her off, and there she stood, with her head and mane up, which looked like a black line down her back; and her pretty black eyes flashed defiance at the pair of us. Indeed I apprehended she would lay hold of me; so, to divert her attention, I touched her with the stick, which she furiously bit through, and then ran off to her young. With this I left her mistress of the field, and felt perfectly delighted at the courage she had displayed in defence of her young.

I made it a daily practice, for about a month, to supply this little heroine with food; at the expiration of which time, as I passed down the hedge-side, there I saw the twelve young ones, and fine fat fellows they were; but no sooner did they espy me than up the bank they ran, showing their little white tails and feet. Through the hedge they rustled into the corn-fields, and I never afterwards saw anything more of them.

CHAPTER IV.

TAME RATS.

INSTANCES of tame rats are by no means rare, or of their becoming gentle and attached to those who feed and caress them. Mr. Bell says, that although the disposition of the rat appears to be naturally very ferocious, still there are instances on record of its evincing considerable attachment not only to its own species but to mankind also; and, no doubt, were not rats held in such universal detestation, the taming of them would be an amusement often indulged in. I have seen numbers of them, at various times, as tame as rabbits; but more especially in the Happy Family, of which I have already spoken, and which may be seen daily in the streets of London. The proprietor will handle and play with them without the least concern, and the little creatures seem quite pleased with his condescension. I called upon him twice, for the purpose of satisfying myself upon two points. The first was, to ascertain if he had any kind of scent about him which might, as it were, charm or stupefy them; and the second was to see if their teeth were perfect; as in the event of their being drawn, that would of course disable them from hurting each other, and they would soon grow tired of quarrelling and fighting.

On both occasions I had a rat and the ferret out in my own hand, and resting on my arms together. The ferret certainly was one of the most clean and handsome animals of the kind I ever saw, and had one of the finest and most perfect set of teeth I ever beheld; and so had the rat. These facts sufficed to show the groundlessness of my suspicions as to there being any drug, charm, or delusion in the matter; for they were as tame and tractable with me as with him, and quietly submitted to every examination without the least discomfiture. Indeed, the only uneasiness they evinced was while struggling with each other as to which should first get into the bosom of my shirt. The rat got in first, and was directly followed by the ferret. In they drew themselves, tails and all, and there they lay quietly snoozing together; so that passers-by did not dream that I had anything of the kind about me. Yet do not suppose that even I was a privileged person, because if you will take the trouble to call, and give the owner a trifle, he will take them out, and they will do the same with you as they did with me, providing you treat them kindly, and not pinch their tails, as some cruel passers-by do.

Here let me warn some thoughtless persons of a wanton act of cruelty. Among the rats in the cage there are several with portions of their tails gone, some having lost half their tails, others nearly all, and so on. On my inquiring the cause, the man told me it was through the spectators cruelly nipping their tails with their thumb-nails, as they popped through the wires; and that where they nipped them, there would their tails rot off. This, I am satisfied will be sufficient to induce any person of feeling to check such wantonness where they see it. But let me proceed with my narrative of tame rats.

In a wild and undisturbed state, how often are rats to be seen so indifferent to man that they will scarcely take the trouble to get out of his way. This indifference arises either from indolence in the man, pressure of business, or kindness of disposition; and thus are rats often charged with daring and impudence which, in truth, is only a confidence they have acquired in man through coming so frequently in contact with him without molestation. I have known instances of their ascending from the bottom of the house to the drawing-room, and eating the crumbs beneath the table that have fallen from supper, while persons were seated at the fireside in comfortable conversation; nor would they go out unless driven; but, upon being left alone, they would clear the carpet, and quietly depart.

In Neale's "Residence at Siam," the author says he was astonished, on visiting the houses of some of the inhabitants, to see a huge rat walking about the room, and crawling up the master's legs in a cool familiar manner. Instead of repulsing it, or evincing any horror or alarm, he took it up in his hands, and fondly caressed it; and then Mr. Neale learned, for the first time, that it was a custom prevalent at Bankok to keep pet rats, which are taken very young, and carefully reared, till they attain a perfectly monstrous size, from good and plentiful feeding. The domestic rats are kept expressly to free the house of other rats; and so ferocious are they in their attacks, that few houses where they are kept are ever annoyed with either mice or rats.

I have met with another instance of the above kind. A friend, by trade a corn-dealer, told me that he had at home one of the finest rats in England, and that he would not take the best ten sovereigns coined for it. Upon further inquiry, he told me that he found it when quite young in the corner of a bin, and that curiosity prompted him to have it emasculated. The consequence was that it grew up one of the finest fellows ever seen, and as tame and playful as a dog. But for keeping the place clear of vermin, it was worth all the cats and rat-catchers in the neighbourhood. For my own part, I feel no hesitation in saying that any one who could feel a fancy for such a thing would find the results most satisfactory.

A man living at Witnesham teaches rats to perform various tricks, such as picking up cards, drinking out of glasses, &c.; and what is even more extraordinary, he has in his possession ten rats, the lightest of which weighs four pounds.

The "Naturalist's Cabinet" gives an interesting account of a gentleman who, about thirty years since, was travelling through Mecklenberg, and was witness to a very singular circumstance:—"In the post house at New Hargard, after dinner, the landlord placed on the floor a large dish of soup, and then gave a loud whistle. Immediately there came into the room a mastiff, a fine Angora cat, an old raven, and a remarkably large rat with a bell about its neck. They all four went to the dish, and, without disturbing each other, fed together; after which the dog, cat, and rat lay before the fire, while the raven hopped about the room. The landlord, after accounting for the familiarity which existed among these animals, informed his guest that the rat was by far the most useful of the four; for the noise he made with his bell had completely freed the house of rats and mice, with which it had been previously seriously infested." This I know, to a certain extent, would have the effect mentioned. But that is not removing the national evil; it is only driving the calamity from your own house into that of your neighbours, and to the country at large. It matters little as to whether the rats have eaten up all farmer Smith's corn, or farmer Johnson's; the corn is missing in the market, and hence is the price of bread affected, without any advantage to the farmer, who has no corn to sell. But I shall mention this subject hereafter, and at present proceed with the taming of rats.

In Lee's "Habits and Instinct of Animals," it is related that two ladies were walking out one day and were accosted by a man who requested them to buy a beautiful little dog which he carried in his arms, and which was covered all over with beautiful, long, curly, white hair. Such things are not uncommon in that part of London, and the ladies passed on without heeding him. He followed, and repeated his intreaties, stating that, as it was the last he had to sell, they should have it at a reasonable price. They looked at the animal, and thought it a most exquisite little creature. The result was that they purchased it. The man took it home, received the money, and leaving the animal in the arms of one of the ladies, went about his business like an honest man. In a short time the imaginary dog, which had been very quiet in spite of a restless bright eye, began to show symptoms of uneasiness ; and, as he ran about the room, he exhibited some unusual movements, which rather alarmed the fair purchasers. At last, to their great dismay, the dog ran squeaking up one of the window-curtains ; so that when the gentleman of the house returned home a few minutes after, he found the ladies in consternation, and right glad to have his assistance. He vigorously seized the animal, took out his penknife, cut off its covering, and displayed a large rat to their astonished eyes, and to its own destruction. But Mr. Lee of course vouches for the truth of this upon the respectability of his authors, and consequently makes apparent two facts—first, that the ladies were perfectly ignorant of the peculiarity of dogs' teeth, &c. ; secondly, that the rat must have been a very tame one to bear so much handling.

It is well known that the Japanese tame rats, and teach them to perform many entertaining tricks, and, thus instructed, they are exhibited as a show for the diversion of the public. Indeed there is no doubt but, through the natural shrewdness of the rat, he might be taught to do many extraordinary feats.

In Belgium, a short time ago, there was a company of theatrical rats, which went through dramatic performances with admirable success. They were dressed up like men and women, walked on their hind legs, and mimicked, with curious exactness, many of the ordinary stage effects. On one point only were they intractable. During the performance the manager had to bring in some food ; but the instant it made its appearance, they forgot their parts, the master, and the audience, and, falling on all fours, set to work most heartily to devour it; that done, the performance concluded by their hanging the stuffed cat, and dancing right merrily round it.

At Rochester, some few years ago, a singular incident occurred. The landlords of the Victualling Office Tavern had a beautiful tortoiseshell cat, the admiration of every one who came to the house. One day she kittened, and all the kittens were drowned. The poor cat felt the loss of her sucklings, and was whining and mewing all over the place in evident distress, but no notice was taken of her. But some few days after, some of the children came across her nest, and saw her in the act of suckling what they thought to be a young kitten. They mentioned the circumstance at the dinner-table, and were laughed at for their trouble; but upon their insisting on the fact, it created some curiosity, and a search was the result, when, to their great surprise, they found the cat suckling, not a kitten, but a young rat! Now it was quite clear that the poor cat had been in extreme pain from an overcharge of milk, and meeting with the young rat, had fondled it up, and from its giving her ease by drawing her milk, her attachment had grown as strong for the rat as for one of her own progeny, which was afterwards manifested to the great delight of numerous customers. If a strange dog came in the house, she would defend the rat with all the vicious determination of her nature; and even after he had grown up a fine, strong fellow, he would, in time of danger, run to her for protection. This curious circumstance spread far and wide, and proved a great attraction to the house; for the rat was as tame as a kitten, and would allow any of the children or customers to nurse and play with it. But, however, to the great sorrow of the landlords and their patrons, a traveller one day called, and, in the absence of the cat, his dog killed poor Master Rat. This was not only a pitiable affair, but a great loss to the landlords; for there is no doubt but many a man has made his fortune by a far less pleasing and remarkable phenomenon in nature.

At the railway stables at Wolverhampton, there was a cat which had a litter of five kittens. Three of the kittens were drowned shortly after their birth, and the cat seemed much distressed at their disappearance. She soon after, however, discovered a rat's nest with a large litter of young ones, upon which she killed the old rat and all the litter but three, which three she carried to her own nest, and suckled them with her own remaining two kittens.

Now this certainly goes far to prove that cats are not such implacable enemies to rats as is generally believed, or they would not, in a state of nature, be so far reconciled, as not only to live with them, but actually, from their own choice, to suckle them in their infancy.

At a tavern in Woolwich, there was to be seen a tame piebald rat, most curiously marked, which was as docile as a puppy, and considered a great natural curiosity.

A friend of mine informed me, that when at home in Edinburgh, he kept a number of rabbits for amusement; and on the floor of the washhouse, where he kept them, lay the remains of an old iron pot which had been used as a copper; but, from some cause or other, a piece had flown out of one side, thus rendering it useless. One morning, when he entered the washhouse, he saw a beautiful and commodious nest of hay, straw, &c., built in the old pot, which was then quite warm. The neatness of the snuggery so excited his wonderment and curiosity, that he resolved upon leaving it alone, thinking he might presently come across the owner. Nor was he far wrong; for on the following morning, on going to feed his rabbits, he looked into the nest, and there lay fast asleep one of the finest rats he ever beheld. He said he could not find in his heart to hurt it, as it was such a beautiful, clean, glossy creature. Presently it awoke, and, instead of showing any alarm or desire to escape, it simply raised its head, and, after looking drowsily at him, opened its mouth, and gaped most lustily; then doubled itself up, and went off to sleep again; thus leaving my friend to mind his own business, and not interfere with him. This off-handed indifference so pleased my informant, that he laid him down some food, and thus left him unmolested to finish his sleep. For some time things went on in this way, till he proceeded, from feeding, to stroking him down the back, and tickling him with his fingers; and in this way was there a perfect friendliness established between them. However, one morning when he went into the washhouse, the rat met him half-way, and with tail erect he reared up on his hind legs, and opened his mouth with such menacing aspect and gesture, that my friend became alarmed, and thought it was time their familiarity should cease; so he ran for the poker, and on his return he found the rat in the iron pot, where he killed him. But he has regretted it ever since, believing, as he now does, that it only wanted to play with him.

A full-grown male rat was caught at an inn in Clerkenwell, and became so tame that the landlord's son used to carry it about in his pocket. It answered to the name of "Tommy," and was very fond of stretching on the rug before the fire along with the cat. For the amusement of the customers, the master would sometimes catch half a dozen mice, and put them into a pan or tub, and then master Tommy would kill them one after the other for the gratification of the lookers-on. His master said he was of infinite value in the cellar, as he used to decoy other rats, sometimes five or six of a night, into the traps, but always avoided them himself, and when his master wanted him it was only necessary to whistle, or call "Tommy, Tommy," and he would instantly come forward, and crawl up his legs to be caressed.

Some time ago the driver of an omnibus was moving some trusses of hay in his hay-loft, when, snugly coiled up in a corner, he found a little miserable-looking rat, whose mamma, having tucked him carefully up in bed, had gone out on a foraging expedition to find something for her darling's supper. The little fellow, being of a remarkably piebald colour, excited the pity of the omnibus driver, who picked him up, and took him home to his family. The children soon took to their little pet, and named him Ikey, after their eldest brother, whose name was Isaac. The little fellow soon grew up, and returned the kindness he had received by excessive tameness and gentleness towards every member of the family. He was, therefore, allowed to roam about the house at perfect liberty. His favourite seat was inside the fender or on the clean white hearth; but, strange to say, he would never get on it unless it was perfectly clean. On one occasion, when the good wife was cleaning the hearth, she gave master Ikey a push; so up he jumped on the hob, and, finding it an agreeable resting-place, there stayed. As the fire grew brighter and brighter, so the hob became warmer and warmer, till at last it became most unpleasantly hot; he would not move from his perch, but rolled over and over, till the hair on his legs and body became quite singed with the heat ; and had they not taken him off, there is no knowing what might have been the consequences. His master held a perfect control over him, and had made, for his especial benefit, a little whip, with which he used to make him sit upon his hind legs in a begging posture when bid, or jump through a whalebone hoop, drag a small cart to which he was harnessed, carry sticks, money, &c., in his mouth, and perform many other amusing tricks. He perfectly understood the use of the whip, for whenever it was produced, and his master's face or voice betrayed anger, in fear and trembling he would scamper up the sides of the room, or up the curtain, and perch himself on the cornice, waiting till a kind word from his master brought him down hopping about and squeaking with delight. In these gambols of mirth he would run round so fast after his tail that it was impossible to tell what the whirling object was, and his master would be forced to pick him up to stop him. At night he would exhibit another cat-like propensity, for he would stretch himself out at full length before the fire on the rug, and seemed vastly to relish this luxurious way of enjoying himself. This love of warmth made him sometimes a troublesome creature, for when he found the fire gone out, and the room becoming cold, he would clamber up gently on to his master's bed, and bury himself under the clothes. He was never allowed to remain there long, if they were awake, but was made to turn out. In that case he would take up his quarters in the folds of his master's clothes, which were placed on a chair; and there he was allowed to remain till the morning. The man became so fond of him, that he taught him, at the word of command, to come into his great-coat pocket. In the morning, when he went out to his daily occupation of driving his omnibus, it was only necessary to say "Come along, Ikey!" and the anxious Ikey was instantly crawling up his legs. He did not carry him all day in his pocket, but put him in the boot of his omnibus, to act as guard to his dinner. But why did not the rat eat his master's dinner? "Because," said the man, "I always gives him his belly-full when I has my own breakfast before starting." The dinner was never touched, except when there happened to be plum-pudding. This Ikey could not resist.

His liking overcame his sense of right, and he invariably nibbled out the plums, leaving the rest for his master. Ikey acted as a famous guard to the provisions; for whenever any of the idle vagabonds, who always lounge about the public-houses where the omnibuses bait, attempted to commit a theft by running off with the bundle out of the boot, he would fly out at them from under the straw; and the villains would run as if his Satanic Majesty were after them; and he thus saved his master's and other property.

The Happy Family.

Having given a variety of well-authenticated facts respecting tame rats, and incidentally adverted to the "Happy Family," a further description of that interesting group may be acceptable to those who have never seen it; and with this I shall wind up the present chapter.

The "Happy Family" are confined in a large cage about six feet by four, and about four and a half feet high. The whole is surrounded by wires; and the vehicle is drawn about like a truck. The interior is plentifully supplied with soft clean straw, and at night illumined with candles, for the sake of public inspection.

In this singular group you see jackdaws, magpies, hawks, owls, starlings, and pigeons—a white cat and five white kittens—six-and-thirty white rats, in addition to others purely black and purely brown; to which may be added a host of piebald young ones of various colours. There are also guinea-pigs, a monkey, and rabbits; and, to crown the whole, there is a magnificent white ferret, and a black-and-white dog. There they all are, snoozing, sleeping, and rolling over each other in one harmonious concord; and nothing in the shape of discord among them. Such a motley group I never saw before; and taking into consideration their opposite natures, some of which are of the most deadly carnivorous character, it was one of the most interesting sights I ever beheld.

The monkey is very kind to his companions; but, like most other monkeys, extremely mischievous. Nevertheless he has formed an extraordinary attachment to one of the young white rats, and is never happy but when it is within sight. I may say with truth, that Jacko has adopted it as his own, for he nurses and fondles over it just as a mother would over her child, and the rat is perfectly conscious of the attachment, and is quite attached to the monkey; so that let the monkey handle it how he may—which sometimes seemed rather roughly—yet the rat never bites him. But, in order to show me the sagacity of the monkey, his master gave him a biscuit, and bade him feed his baby. He immediately caught his favourite, and, placing it in his lap, gave it a piece, and then had a mouthful himself; yet he had a great objection to the rat having more than its share, which, to tell the truth, was sometimes a very small one.

I have watched this Happy Family for hours together, and all is one unchequered scene of harmony, except now and then, when the monkey, who is king of the colony, is taken with fits of mischief. For instance, when they are all embedded in one corner, and fast asleep, he becomes lonely and unhappy. Down he will jump, and, like a peevish old bachelor, in the bottom of a lumber-cupboard, seeking his lost slipper, he commences groping about for his favourite; and, should he not at once meet with it, he shows his royal indignation by seizing the kittens, rats, ferret, and guinea-pigs by their heads, tails, backs, or bellies. Away he sends them, right and left, flying in all directions to the other end of the cage; but when he finds his favourite his anger ceases. Indeed he is never quiet. Sometimes he will roll his pet on its back, and, with all the anxiety of an affectionate parent, will turn up the fur with one hand, and catch the fleas with the other,—a job he is very fond of, and to which parental solicitude the rat yields with all the complacency of a little fat baby. There is no trouble in finding out which is the monkey's favourite, for its fur is all turned the wrong way with rough nursing, which makes it look more like a little white hedgehog than a rat. At other times his grotesque majesty will take an instantaneous tour through his dominions. Away he flies, with the rapidity of lightning, all over the cage; and then, bounding from side to side, wantonly sweeps the perches as he passes, upsetting hawks, owls, jackdaws, magpies, starlings, and pigeons, and pitching all the animals that come in his way up to the ceiling; so that, with the fluttering of birds, and the helpless flight of cats, rats, ferret, and guinea-pigs to the ceiling and back again, the cage appears crammed with fluff and feathers; and it is a question whether the great earthquake at Lisbon caused a more instantaneous consternation than does his bobtailed majesty in the Happy Family, when seized with his periodical propensity for polking.

CHAPTER V.

RATS' NESTS, AND THE MATERIALS FOR BUILDING THEM.

THE female rats about the water-side, or in the fields, &c., select a secluded place where the male rats are not likely to find them, and there build their nests with soft dried grass, leaves, wool, or anything else that chance may throw in their way, and which may tend to make it soft and comfortable. Here they give birth to a naked blind family, which they watch and attend to with all the fondness of the most kind and affectionate of parents. But the rats about farm-houses and towns seem to spurn such simple rusticities; since, for the most part, their nests must be built with silken, linen, woollen, or cotton fabrics, and lined with paper, feathers, or furs. They are not very particular; for anything will suit them, from a lady's cambric nightcap, silk stocking, Bloomers, Brussels lace, or sable tippet, down to the coarsest brown paper or dried monkey's skin.

Among the furtive propensities of rats for obtaining their building materials, I give the following instances:—

A gentleman I knew, and his domestics were continually missing various small articles of both male and female wearing-apparel, such as socks or stockings from their bedsides, which they had pulled off only the night before, and also cambric collars, neckties, silk handkerchiefs, or napkins; in fact, any small article which was easy of conveyance. These continual losses of course led to numerous suspicions, which of necessity must have been the cause of many nights and days of uneasiness to the whole household; for, in a large miscellaneous business, where all articles of merchandise are constantly exposed, and where moneys are continually changing hands, what can be so cruelly harassing as to have clear and undeniable cause of suspicion that there is some dishonest person about the establishment. Then who could be the thief? there was the question! One thing was certain, that he who could find in his heart to steal a man's silk handkerchief would feel few qualms of conscience in stealing his purse, if an opportunity offered. But, on the other hand, what could they want with a single sock or stocking, leaving the fellow one behind? There was the paradox. Yet, upon a more extended view of the matter, it was quite clear that the felonies were not perpetrated by any outdoor thieves; for, had they made an entrance at any time, they would not have taken such minor articles as silk handkerchiefs, odd socks, stockings, or napkins, when there were articles of far more serious value at hand. Besides there were no doors, shutters, or bolts displaced. No, no; the thieves were in the house; and thus things went on for months and months, and, in despite of every care and precaution, still things were continually missing, till their hopes of discovery fell into despair, and gave birth to the most reckless and perfect indifference.

Some time after the master of the house resolved upon having his store-room enlarged, in the corner of which there was a formidable copper, which was connected by means of a flue to a large projecting chimney-stack. All these he resolved upon having removed, as it would render the kitchen so much more roomy and convenient. The bricklayer was sent for, and scarce had he received his orders when he set to work to knock down the flue, and out tumbled eleven young rats, besides a host of old nests which, from time to time, had been built by the rats, till the flue was completely choked up. The copper had not been used for years; but when the bricklayer brought all the old nests to daylight, to the great joy of every one in the house, there were all the missing articles; and a strange motley group they presented. There were pieces of old stockings, old shirts and towels, sundry pieces of silk, flannel, cotton, and woollen cloth. Besides these, there were the remnants of various silk handkerchiefs, gloves, and a host of other things.

In a cellar, in Holborn, there existed a colony of rats,

which, from being seldom molested, had eventually become so daring and bold that they even disputed the right of ownership with regard to the food that people might carry upon their persons. One day, the dried skin of a monkey, which had been suspended from a nail in the cellar, preparatory to its being stuffed and preserved, was suddenly missed, when, upon examining the premises, it was discovered behind some loose lumber, cut and torn into a hundred remnants, and fashioned into a commodious nest, in which were encradled five young rats. The infantine progeny were summarily destroyed; but the much-valued skin of the monkey had been rendered entirely useless by the mischievous and destructive parents.

A lady, residing at Malvern, had occasion to leave her home for a time; but on returning, after three months' absence, she fancied her pianoforte sounded very curiously, and, on sending for a tuner, he discovered that a rat had gnawed a hole through the bottom, and taken up his residence in the interior, where he had built himself a commodious nest with the leather coverings of the hammers, with some portions of silk, and other articles of delicate composition, which he had foraged from the establishment.

The Bloomers and the Rats.

In a retired spot in the neighbourhood of London there resides an elderly couple, who, in the downhill of life, are enjoying a comfortable independence. They have only one child, an amiable and interesting daughter, by the name of Eliza. They originally came from Yorkshire, and possess the peculiar qualities of the humbler class of that county to their fullest extent, namely, an unqualified respect for themselves and for good living. They keep no company, for two ostensible reasons: first, because they never learned to read or write, and consequently are utterly disqualified for superior society; and secondly, because their independent means have elevated their notions far above anything less than carriage acquaintance. In this retired establishment the daughter is the only one that can read, and is of course the oracle of the house. It happened that a few years back she read a great deal in the newspapers about *Bloomerism,* and the more she read and reflected, the more she became infatuated with the new American costume. At last she resolved upon adopting it, in order to cut a dash before her country cousins. The matter was soon broken to her parents, who as speedily fell into her views, being themselves naturally fond of anything showy and uncommon, and Bloomerism became the order of the day. A first-rate dressmaker was immediately sent for, by whom she was completely equipped, in the most costly style,—making altogether a sum total of £20. 11s. 6d. Thus the charming girl, to the delight of her parents, appeared a first-rate Bloomer; and when the feast of admiration was over, the old gentleman insisted upon her going directly to her uncle's, and staying the night, when her mother should fetch her home in the morning. No sooner said than done! The Irish servant, in her enthusiasm, without either cap or bonnet, ran through the streets, to bring a cab; and as both residences lay close to the railroad-stations, one hour brought her to her uncle's farmhouse. Her aunt and cousins, two amiable, healthy country girls, seemed perfectly paralyzed with astonishment; nor did their distended eyelids attempt to wink till she assured them, on the word of a lady, that she was their cousin, Eliza, from London. "Why, hang it, lass," said her uncle, "beest it thee? Who has dressed thee in a jacket and breeks? Thee looks just like a player chap, as I seed in front of a show at Bartlemy Fair!" "O uncle," said Eliza, "this is my new Bloomer dress; it's all the fashion now among gentlefolks!" "Well, thank God," said he, "we're not gentlefolks! But come in, an' let's know how your father and mother are, and how Lunnun goes on."

The evening was spent in sipping ale and cracking jokes, till bed-time was announced, when all retired to rest—Eliza, of course, to sleep with her cousins, as there was plenty of room, the bedstead being a large four-poster. After many a little conversation, and merriment, the light was put out, and in this retired spot all was dead silence. They had not been in bed many minutes, when there was a patter, patter, patter, followed by a bump,—after that came more patter, patters, when Eliza, in a subdued whisper, asked her cousins what it was; they told her it was only the "rots," and that she was not to be afraid, for they wouldn't hurt her! Now

Eliza was terrified at the very name of them; so she quietly drew the clothes over her head, and crouching close up to her cousins, lay as if dead. The younger cousin, to satisfy her, put her hand out of bed, and picking up one of her boots, threw it in the direction of the noise, and, it is thought, knocked down the Bloomer hat; however, for a few minutes all was still again, which gave poor Eliza time to breathe. Presently the noise returned, followed by a second, and a third; after them came plenty more; and then there was a clatter, clatter, which preceded a hurry, scurry, interspersed with sundry lumps and bumps, which sounded as if they were playing with cobbler's lapstones. Then came other hideous sounds, till at last you would have thought there were at least a dozen coach-horses dancing a polka. In the midst of all, then commenced a most terrific battle, which ultimately proved to be a contest for the Bloomers; and to such a pitch of desperation did they carry the conflict, that the two cousins, for the first time in their lives, were literally scared; and, like poor Eliza, tremblingly hid their heads beneath the clothes to shield them from these daring burglars. There they lay steaming and quaking, when their father, from the bottom of the stairs, called out to know what noise that was; and, not receiving an answer, he called out three times; but they were too terrified to answer, or even put their heads out, hence he concluded they were fast asleep, and returned to his bed. But his calling out startled all the rats; nor did they return, at least while the girls were awake; for there they lay listening and listening, till every sense grew dim and weary, when Morpheus, entwining them in his leaden arms, at length lulled them to sleep.

The following morning, the dame was up betimes to get her husband's breakfast ready, and put everything in prim order for the comfort of her guest. Hour after hour rolled on, till the clock struck ten, when the good man returned for his luncheon.—"What! the girls not stirring yet, dame?" "No, she had heard nothing of them; but she supposed they were tired!" "Well, just get us a jug of ale, an' then go an' wake 'em up, or they'll sleep all day, mon." The ale was soon on the table, and up to their bedroom she went and knocked, but received no answer; for, what with their being kept awake with rats, and the nervous terror they had experienced, assisted by the narcotic effects of the carbonic acid gas they were inhaling beneath the clothes, there they lay, enfolded in each other's arms, like sleeping graces in a group of statuary! Dame knocked again, when a faint voice said, "Come in." In she went, and oh, what a picture of horrors presented itself! She shrieked out, "Oh, look'ee here! whatever shall we do?" The three girls shot up like a Jack-in-the-box, and sat bolt upright in bed, with their eyes and mouths wide open. "Dear, dear, dear!" said dame, "whatever shall we do? Here's this beautiful velvet jacket on the ground, and covered all over with dirt—we shall all go mad—an' there's the beautiful petticoat just as bad. I never saw anything so cruel in all my life—an' there's the beautiful hat thrust into the corner without its feather—but where are the Bloomers gone?"

Here the two daughters, assisted by their mother, sent forth such shrieks of lamentation and horror, that it aroused the good man from his luncheon; and as for poor Eliza, she fell back in hysterics. In came master: "Why, what i' the world's the matter?" "Matter, John," said dame, "just look'ee here;—I'll lay my life on't, 'tis those cursed rats!" John scratched his head, and with a vacant gaze, said, "Very like, very like."

Till now, they had not noticed poor Eliza in hysterics! "Dear, dear," said dame, "here's a peck o' trouble; the poor child's fainted away." Downstairs she ran for some little antidote, and met Eliza's mother at the bottom, who had just arrived from London. John, in the interim, went to pick up the hat, and saw in it a large rat and a number of young ones. He roared out for his dog Boxer, and at the same time kicked at the hat with all his might, and knocked it into all manner of shapes; when in rushed Boxer, "Rat, rat! boy!" said master, urging on the dog. But it would appear that he had killed the old one and some of its young with his kicks. The dog thrust his muzzle into the hat, and killed the remainder, and then became so excited, that he seized the hat in his jaws, and shook it till it rattled again, when out flew all the dead rats and pieces of ostrich-feather; for the old rat had bit the feather in pieces to make a soft nest for her young. At last Boxer finished his work by fixing his fore feet upon the hat, and tearing it all to pieces, just as the two mothers entered.

To describe the old lady's horror and indignation at the scene were a thing impossible. Besides, always when she came down to the farm, it was her invariable custom to put on her utmost dignity, because master was her younger brother, with whom she always exercised the right of quarrelling; consequently they never met without a row.

By this time Eliza had somewhat recovered; and John, to evade his sister's glances, as well as to find out where the rats had their retreats, went groping into all the corners, and lastly into an old cupboard. "Holloa," said he, "why, what's this down the rat-holes?" at the same time dragging it forth, and holding it up to view,—"Why, dang my buttons," he roared out, "if 'tisn't the gal's Bloomers!"

It would appear that, in the contention overnight, the rats had slit them up to the waistband; and, in their retreat, each party had dragged a leg down their respective holes. You may easily imagine what state the Bloomers were in. The old lady, in a towering passion, rushed at John, and snatching them out of his hands, declared he was "the most ignorantest fool she ever knowed;" and concluded by stating, what a curse it was to think that she should be related to such a brute.

The old lady was gathering herself up for a perfect storm, when John told his wife that he could be of no further use to them, so he would go about his farm, and leave them to settle the matter the best way they could.

And now I shall conclude with two more observations: first, that this affair had the effect of entirely curing Eliza's propensity for Bloomerism; and secondly, that the cause of these disasters seems to have arisen from the fact, that Eliza, being out for a day's holiday, had used musk rather freely. Rats, being naturally fond of that scent, had been attracted by its odour to the spot; and hence arose the great battle for the Bloomers.

CHAPTER VI.

DIETETICS OF RATS.

We now come to that part of *Ratology* which most concerns the interests and well-being of the human family, namely, the rat's stomach.

Of all animal stomachs I believe the rat to possess the most astounding and convenient one; for it can adapt the intestines to every kind of digestible substance that chance or locality produces. Rats will eat all kinds of grain or farinaceous food, from a sago or tapioca pudding, hot or cold, with or without sauce, down to horse-beans, peas, or coarsest barley or pea meal, including all kinds of pastry, from the choicest cheese-cakes or custards down to the commonest hot or cold cross-bun or sailor's biscuit.

As for fish, rats have no mercy on them. They will eat all kinds of small fry, heads, tails, bodies, bowels, bones and all, from the delicate whitebait or smelt down to the rusty red-herring of Scotland. Indeed, nothing comes amiss to them, from the whale to the shrimp. Hence arises the cause of their locating, during the summer months, at the water-side.

In the "Sporting Magazine" there is a grave series of charges brought by a gentleman against rats for their depredations among corn, game, and fish. In speaking of the last, he says: "There is another and most serious evil occasioned by rats; that is, the destruction of fish in streams, pools, and stews, where fish are preserved." He says few persons have any notion of the quantity of rats that frequent these places. He does not mean the more innocent, brown, short-eared, small, bright-eyed, and pretty-looking water-vole, but the coarse, fierce, grey when old, Norway, farmer's rat, the frequenter of houses, buildings, ricks, hedges, plantations, brooks, pools, and every place under the heavens; while the vole, neither in winter nor summer, ever deserts the water, but on a summer's evening may be seen quietly seated by the side of the stream, munching the white root of the bulrush, of which it is particularly fond, and holding it up in its paws, squirrel-fashion, seems to enjoy its evening repast as much as an alderman does his whitebait dinner, if that were possible.

The same writer furthermore states that the Norway rat, in the summer months, frequents the water, and will attack a large fish in shoal water, and soon master it. Besides, he has known them come out of their holes, and carry away six or seven fine perch, which had been caught and left by the pool side, with the greatest ease :—

"Some time ago my son had just returned from a day's angling at Hanwell. The fish lay sparkling on a dish for my approbation. He had caught ten, but he informs me that the first three he caught were by far the finest; and in order to have them safe, he threw them on to the grass some few yards behind where he stood; then, after standing for some time quietly watching his float, he casually turned his head, and there were two large rats running away with two of his fish. He directly dropped his rod and pursued them, but they reached the water before him, and dashed in, taking the fish with them, and instantly disappeared. He supposes that one rat first found them, and taking away a fish to his hole, brought his companion to help him with the other two, and so the three finest fish were lost."

A gentleman was walking alongside a millstream near Newcastle-upon-Tyne, and noticed a common house-rat making its way close by the edge of the water, among the coarse stones that form the embankment. Curious to know what it could be doing there, he watched its progress downwards, until it reached the outlet of a drain. It had scarcely turned into the drain when it made a sudden plunge into the water, and almost as quickly reappeared in the stream with a middle-sized eel in its mouth. It made for the edge, where it regained its footing, and this, from the steepness of the bank, was a matter of great difficulty, which was much increased by the struggles of the eel to get free. Eels at any time, as every angler knows, are troublesome gentry, and very hard to manage; consequently would require all the ingenuity of a rat to cast a knot on one's tail. But when the rat attempted to get forward, and turn a corner where there was a broader ledge, the desperate efforts of the eel rendered his footing so precarious that, rather than have a second plunge for it, he was reluctantly obliged to drop it into the water. His first action afterwards was to give himself a good shaking, both to revive his spirits and to rid his coat from the effects of his morning dip ; and then, as before, he resumed his fishing recreation till he got out of sight,—the stream preventing the observer from following him further.

As some labourers were cutting through an embankment in a field adjoining the river Lune, they met with between fifteen and twenty pounds' weight of eels, some quite fresh, and others in the last stage of putrefaction. They varied from a quarter to half a pound each, and consisted of the common silver-bellied or river eel, and Liliputian specimens of the conger or sea eel. The latter of course had come up with the tide. As teeth-marks were visible on the heads of most of them, it was conjectured they had been destroyed in that way and stored for winter provisions by some animal whose retreat was not far distant. This proved to be the case, for, on digging a little farther, out bounced a matronly rat with seven half-grown young ones at her heels. The workmen gave chase, and ultimately succeeded in killing both mother and young ones. The embankment is about a hundred yards from the water's edge ; so that it must have cost considerable time and labour on the part of old Ratty to catch and drag the eels thither.

Rats swarm about the small towns in Scotland where the herrings are cured, living amongst the stones of the harbours and rocks on the shores, and issuing out in great numbers, towards nightfall, to feed on the stinking remains of the fish. At the end of the fishing season they may be seen migrating from these places in compact bodies, and in immense numbers. They then spread themselves, like an invading host, among the farms, farm-houses, and stack-yards in the neighbourhood. They again repair to the coast for the benefit of a fish diet and sea air ; their wonderful instinct telling them that the fishing season has again commenced.

In the fish-markets of London, and also in the lower order of streets, where fishwomen are in the habit of standing, rats have from time to time been seen issuing forth, after midnight, to eat up the heads and entrails of fish, which the day's sale had left. Thus before scavengers were introduced, they were of infinite benefit, though their services are now no longer required for that purpose. But about slaughter-houses, knackers' yards, victualling depôts, drains, &c., their capacious stomachs are still of inestimable value to the population, by consuming all kinds of animal and vegetable refuse, that would otherwise be left in the drains to putrefy, to the great danger of the public health. As to animal substances, rats will gloat over and devour anything, from a delicate chop of house-fed lamb, or babies' fingers, down to a venison pasty, an old tortoise, or putrid carrion. But with respect to poultry and game of every description, nothing dead or alive, either on water or land, is safe from their rapacity. They will eat anything, from the delicate wing of a roast duckling, young partridge or pheasant, down to the scaly old leg of a centenarian swan. They will likewise consume all kinds of oils and fatty substances, from the purest olive oil to the refuse of whale's blubber. Nor will they object to soaps, either yellow or mottled, tallow fresh or stale; nor are they very particular, in times of need, as to boots and shoes, or horses' harness. They will also consume all kinds of tuberous or bulbous roots, from a prize tulip to a mangel-wurzel. I have read also of their getting into churchyards, and eating our departed friends in their graves, as well as infesting the dead-houses on the Continent, where the bodies of strangers or casual dead are taken, for the purpose of being owned and claimed by their friends; but frequently, in a single night, their faces and portions of their bodies have been so completely eaten away by rats, that all traces of identity were entirely obliterated. Thus it appears we are never secure, either dead or alive, from the liability of becoming food for rats.

Notwithstanding the weak and contemptible appearance of the rat, it possesses peculiarities and properties which render it a far more formidable enemy to mankind than even those animals gifted with the greatest strength and most destructive dispositions, such as lions, tigers, wolves, wild cats, hogs, and hyænas. The midnight burglaries undetected by the police sink into in significance compared with the ravages of the rats of the London sewers, which steal and destroy more in one week than the value of all the robberies of plate that blaze away in the newspapers from one year's end to another. They are one of the greatest animal nuisances that have infested our homes and fields since the days when an English king levied tribute of wolves' heads upon our brethren of Wales.

Independently of their destroying furniture, &c., they have been known to gnaw the extremities of children while asleep. A child was nearly eaten to death by rats in the City. The parents, it appears, lived on a first floor, and the mother had gone out to market, leaving the child alone, sleeping in the cradle. During her absence the persons on the ground-floor heard the child crying in a most piteous manner, and after some time they went up to see what was the matter. Upon entering the room, they beheld several rats gnawing one of the child's hands, two fingers of which they had actually eaten off. The child was immediately taken to the hospital, and had the lacerated parts cut away, and fortunately no fatal consequences ensued. The rats are supposed to have effected their entrance from the drain underneath the house communicating with the main sewer, and, but for the timely interference of the occupiers of the ground-floor, there is little doubt the child would have been entirely eaten up. The circumstance at the time occasioned considerable sensation in the neighbourhood; but, like all other rat-exploits, was soon looked upon as a mere matter of course, and then sank into comparative oblivion.

One evening, as a gentleman well known to the theatrical world was seated with his family at the supper-table, they were all at once dreadfully alarmed by the heart-rending and pitiable screeches of his infant daughter, who had been sleeping in the adjoining room. They instantly ran to ascertain the cause of her agonies. At first they saw no visible cause, but on slightly turning down the bedclothes they discovered, to their horror, that blood was streaming from one of her feet, and upon closer examination they found the joint of her great toe most dreadfully lacerated. Of course, medical assistance was immediately sent for, and in the interim their imaginations were strained to their utmost as to how or what could have been the cause of it. While thus pondering, they suddenly saw something moving backwards and forwards beneath the clothes, at the bottom of the bed. The first impulse of the father was, of course, to grasp at it outside the clothes, and squeeze it with all his might; this he did, and held it till it was dead. Then, upon throwing off the bedclothes, they beheld, to their loathing and disgust, an enormous sewer rat. When the medical gentleman arrived, and saw the injured foot, and also ascertained the cause, he resorted to such means as would purify the wound from all poisonous effects, by well cleansing, &c., which happily terminated in nothing more serious than her being crippled in that foot for some time, and wearing the scar as a remembrance. But its mother informed me, that there was no doubt, from the desperate wound inflicted, that had they not instantly run to the child's rescue, the rat would soon have had her toe off, if nothing worse. This occurrence took place some time since. But now we come to others which have transpired within more recent periods, and of a more dreadful description.

The wife of a labourer, residing near Landor, went out of her cottage, leaving her infant boy, about three years old, asleep on the bed. On her return she heard the child crying vehemently, and upon rushing into the room she saw a large rat busily engaged in biting the little fellow's face; but on her appearance the animal ran up the chimney. It was found that portions of the child's flesh had been eaten away, both from the face and one of its thumbs.

The death of an infant six months old took place in Marsh Street, Bristol, from the bite of a rat. Marsh Street is an exceedingly old and filthy street, and lies not only adjacent to the floating harbour, but in the vicinity of numerous warehouses for the storage of potatoes, grain, tallow, oil, and the like; and, as may be expected, many of the houses are infested by rats. The mother, on hearing her infant scream, hastened to ascertain the cause, when she found that her poor child had been attacked in a most ferocious manner. The rat had severely bitten it under the right eye, and marks of the creature's claws were still visible on its face and neck. There was also a great quantity of blood about the person and clothes of the victim. The mother paid the poor little sufferer all the attention in her power, and a medical man was consulted, but it ultimately died from the injuries it had sustained.

A few years ago, the town of Dowlais was the scene of a most painful and revolting occurrence. Some of the poorer class of houses are infested, to a considerable degree, by rats. A poor working woman having occasion to go from home, put her infant child to bed. Upon her return, and opening the door of the apartment in which her infant lay, she saw three large rats jump from the bed, and, on looking in the direction of her child, she was terrified at perceiving that the bedclothes were stained with blood. She instantly removed the coverlet, when a shocking spectacle presented itself. The rats had mutilated the poor infant and destroyed its life, having eaten away the wall of the belly, and actually destroyed portions of the intestines.

I shall conclude this calendar of infant sufferings and mutilations with one case more, which took place in Dublin, and which is, if possible, more appalling than all the rest. From the testimony of the unhappy mother of the child, which was given on the coroner's inquest, it appeared that she had committed it to the care of a woman; and it was whilst under this woman's care that the infant received the injuries which caused its death. Her evidence was to the effect, that on the night in question she fed the child and placed her in the cradle to sleep. She was awoke in the night by the child screaming. Witness got up, and quieted the child, and she went to sleep again. In the morning, at seven o'clock, witness got up, and, on approaching the cradle, found the child and the clothes about her all over blood. On her lifting the clothes off the cradle, two huge rats jumped out, and ran under the bed. She immediately ran with it to the hospital. According to the evidence of the surgeon, the child, when brought into the hospital, was fast sinking from the loss of blood, and half the inside of the left hand was eaten away, and the right arm was frightfully gnawed, evidently by rats; the face was also torn. Despite of every care, the child sank, and expired that morning from the injuries she had received.

It appears that children are not the only victims, when

rats are seized with a craving for human flesh. The following circumstance has been mentioned by various authors. There was a German bishop, by the name of Hatto, whose residence was infested with so many of these animals, that he built a tower, close to the Rhine, for his defence against them. Here he resided ; but at last they gained an entrance, and at length killed and ate him.

A few years ago a friend of mine was in Dublin. There was a tradition current, to the effect that some time previously a British officer had come by his death in a most melancholy manner through rats. The account ran as follows. A bosom friend and brother officer of his had died of a fever, and he among others attended the funeral. When the ceremonies were over, and all the mourners had retired, he sought an opportunity of leaving the company, and went alone into the vault to pay a last tribute to his departed friend. Now whether, according to the custom of the country, he had partaken a little too freely of whisky, and therefore fallen asleep, or whether he was so completely absorbed in devotional supplications for the welfare of the soul departed, nothing has transpired to determine. But suffice it to say, that in the evening the gravedigger came, and never supposing for a moment that any one was there, closed up the entrance of the vault, and so fastened him in. On the following morning, upon the soldiers gathering, and the muster-roll being called, he was found absent, and being one of the most regular in his attendance, it caused an inquiry as to who had last seen him ; but no one had set eyes on him since the evening before. This caused some uneasiness among his friends, since they knew of his devoted attachment to the deceased officer. They called at his lodgings, and ascertained that he had not been home all night. That caused a hue and cry, when the thought suggested itself to some of his friends to have the vault searched. The gravedigger was soon sought for and found. On opening the vault, there lay the missing officer a corpse, and so miserably gnawed and mangled by rats, that, but for his uniform, they could scarcely have identified him. But on further examination some dead rats were found, and were supposed to have been killed by random cuts with his sword in the dark, since that instrument is said to have been found by his side besmeared and clotted with blood and fur. Now whether he died from fright or exhaustion, or from both, or whether the rats in a body followed him up and killed him, will remain to all time a mystery.

About forty years ago there was a pie-maker of Leith who met with his death through rats. It appears that he had gained great repute for his peculiar meat-pies, which were considered both good and cheap. He was an industrious man, and acquired a large connection. It was his invariable custom to make his meat-pies over-night, so as to be ready for the morning's baking. But for some time past he had found the meat taken out of the pies in the night-time. Thus matters went on till pies and all went. This so puzzled him, that he resolved upon making his pies as usual, and stationing himself in the bakehouse in the dark to watch for the thieves. This he did. On the following morning the neighbours were stirring, but his shop remained shut; they concluded he had been awake all night, and was having a morning's nap, and therefore would not disturb him. Midday came, and still his shop was shut. At last his customers began to muster for their pies; and sundry inquiries were made as to what had become of the pieman. At last they burst open the door; still no pieman made his appearance. They then searched the house all over, but to no purpose. At last they went into the bakehouse, and there they found him on the floor, a corpse, with his face, hands, and body most frightfully gnawed away. The bystanders were perfectly horror-stricken, and could form no idea as to what could have so cruelly maltreated him. At length one of them passed out into the yard, and there lay a large wooden pump which had belonged to an old ship. He saw a rat pop out at the further end, which, as soon as it saw him, popped back again. This threw some light upon the mystery, and quietly searching, they found that where one end of the pump came against the outside of the bakehouse the rats had made a hole, and so got access to the interior of the bakehouse. They then set quietly to work, and stopped both ends of the pump tight up, and so fastened all the rats tight in. Then all the pots and kettles in the neighbourhood were set to work for boiling water, and after boring sundry holes in the upper side of the pump, they poured in the boiling liquid, one kettle after another, till they scalded all the rats to death; and upon turning them out, it is said, there were hundreds. Now whether the pieman, like the young woman of Paris, died from fright, and was gnawed before he was cold, or whether the rats attacked him in a body and killed him, will remain to the end a mystery.

There lived lately in Clerkenwell a woman aged 63, of rather eccentric habits. Not having been seen for some days by her neighbours, suspicion was aroused, and upon her room being entered, her lifeless and putrefied corpse was discovered lying in the middle of the apartment; the face, neck, and other parts of the unfortunate woman's person having been so awfully mutilated by rats as to be hardly recognizable.

From many striking instances, it appears very clear that rats are creatures of impulse, for frequently they will pass by a thing without caring for, or even noticing it, when at another time they will seize upon it with the most ferocious daring, and devour it with avidity; nor are they very particular, in the absence of anything tender, at trying their teeth with something tough. For instance, the author of "Gleanings of Natural History" gives an account of an old tortoise. But before I relate this circumstance I may as well give an account of the rats of the Cape of Good Hope.

A gentleman residing in London, wrote to the "Gentleman's Magazine" to the following effect:—"In the course of his life, public service had carried him several times to the Cape of Good Hope, where it struck him as a strange fancy in every family, to see a small land-tortoise in the inclosed yard, behind the offices of the house. For some time he looked upon the animal as a universal pet; but at length he was undeceived, by being told that they were kept for the purpose of keeping away the rats, which would never approach any place where a land-tortoise was harboured."

So much for the modesty of the Cape rats; and I might say also the rats of Calcutta, for Mr. Jesse says, that, while on the tortoise, he may as well mention that Captain Gooch informed him that when he was at Calcutta, he was told that a tortoise which had belonged to, and had been a great favourite of, Lord Clive, when he was Governor-General of India, was still living. He went to see it, and as no one seemed to take any interest in the creature, he procured it with little difficulty, and brought it to England. But before he left Calcutta he made every inquiry as to the probable age of this tortoise, and ascertained, from a variety of corroborative circumstances, that it could not be less than two hundred years old. On his arrival in England, Captain Gooch had the old tortoise put into the coach-house, at his seat near Clapham Common. There for a short time it did well; but one morning, nothing was found of it but its shell, the poor old tortoise having been killed in the night, and devoured by rats.

Mr. Jesse relates an extraordinary instance of the sagacity and foresight of rats, and, wonderful as it may appear, he says it may be relied upon, for he received it from a person of the strictest veracity, who was an eye-witness to the fact. A box containing some bottles of Florence oil was placed in a store-room which was seldom opened : the box had no lid to it. On going to the room one day for one of the bottles, the pieces of bladder and cotton which were at the mouth of each bottle had disappeared, and a considerable quantity of the contents of the bottles had been consumed. The circumstance having excited surprise, a few bottles were filled with oil, and the mouths of them secured as before. Next morning the coverings of the bottles had been removed, and some of the oil was gone. However, upon watching the room, which was done through a little window, some rats were seen to get into the box, and insert their tails into the necks of the bottles; and then withdrawing them, they licked off the oil which adhered to them.

A friend of the author's lately received from a kind individual, resident in a foreign country, a package containing a few bottles of salad oil of the most delicious kind, a present which, to say nothing of the respect that was shown to him, afforded him unspeakable delight, being an ardent admirer of this sort of dainty. The bottles, which were carefully sealed, were placed in the apartment allotted for eatables, there to lie safely, as was supposed, till wanted. A few days after, our friend had a great desire to have another look at the present, and so betook himself to the pantry, when he found, to his surprise and indignation, that a nice little hole had been made in all the bungs, and some of the contents extracted from each of the bottles. There was something incomprehensible to him about the matter. He could understand how a cork or bung might be eaten through by rats or mice, but how they could manage to get at the contents was a mystery, the hole being too small to admit the head of either of these animals. Determined to ascertain who the delinquents were, and the means used by them to effect their purpose, he secreted himself one night in a corner of the room, and soon a fine glossy rat made its appearance; approached the box with a fortitude unknown to rational depredators on a similar errand, poked his tail into one of the bottles, then drew it gently forth and licked it clean, and so repeated the process over and over again till he had had his fill.

At a place in the neighbourhood of Manchester, where a station has been abandoned and the old wooden hut removed, a gentleman lately saw an ingenious and novel theft committed, and allowed it to be completed without molesting the robber. He happened to be standing quietly by, when he saw a fine sleek rat come from beneath the old station office, and walking deliberately up to a carriage that was standing off the line, clambered up a spoke in one of the wheels to the box wherein the grease was kept; and, as if regularly trained to the office, with one of his fore-paws he raised the spring-lid, and there held it while he looked round to see if any enemy was at hand; then, seeming satisfied that all was safe, he forthwith plunged his nose and whiskers into the grease, which we believe is composed of palm-oil and tallow, and which he seemed to eat with as much relish as an alderman would the green fat of a turtle. But from time to time he drew out his head to see that all was right, still holding up the spring-lid with his paw, and as he felt satisfied that all was secure, he again plunged into the grease, and so on till he had had his fill; after which he let fall the lid, and quietly and steadily returned to his abode. This rat, I think, showed as much cunning and sagacity as the rats that put their tails into the oil-flasks; but still it is a matter that will admit of discussion.

I have seen and nursed the wonderful little dog Tiny, which, for its size, was the greatest rat-killer the world ever produced; at least, we have no records of any dog coming near it. He used to lie for exhibition on a crimson cushion, placed upon the table in the bar-parlour, with mould candles on either side, so that the customers and the curious could see him from the front of the bar. Just above where he lay there stood a shelf; and upon this shelf, the wife of the owner of the dog, on one occasion placed a paper bag containing four pounds of lump-sugar. In a day or two there was a call for sugar; away she ran to reach down the four pounds; when, to her utter astonishment, the bag was empty. She, however, soon discovered the cause; a rat had drilled a hole through the wainscoting just behind the bag, and thus carried away every lump of the sugar. This certainly was a theft of great daring on the part of Master Rat, considering it was perpetrated within a yard of the notorious Tiny, the great enemy and destroyer of the rat tribe.

Now that we have seen their liking for sugar in its most refined state, let us see what relish they have for saccharine, or, more plainly speaking, sugar in its rawest state.

In the "Natural History of British and Foreign Quadrupeds," the author tells us that a gentleman had an estate in Jamaica much infested with the native rat, which I have been informed is a very pretty little animal, but to which he had a great dislike; and, as the author says, he imported, at great cost and trouble, a large and strong species to exterminate them. The rats he imported were, I suppose, our common brown rats, as I know of no others that would have answered his purpose so effectually, or carried matters to the extent which it appears they have done.

We are told that these rats went far beyond his expectations or wishes; for, after disposing of the native rats, they extended their hostility to the cats, and killed them also; and thus got rid of two enemies at once.

In no country is there a creature so destructive of property as the rat is in Jamaica—their ravages are inconceivable. One year with another, it is calculated that they destroy at least a twentieth part of the sugar-canes throughout the island; but this is not all—they prey upon the Indian corn, and on all the fruits within their reach, as also roots of various kinds, and indeed anything that is digestible. Some idea may be formed of their immense swarms from the fact, that on a single plantation no less a number than 30,000 were destoyed in one year. Traps of various kinds are set to catch them ; poison is sometimes resorted to for killing them ; as also terriers and ferrets sometimes to hunt them out ; nevertheless their numbers seem undiminished, so far at least as can be judged from the ravages they commit.

It is a fact well authenticated, that rats prefer mild Wiltshire breakfast bacon, to lean, salt Irish ; and as for hams, in the absence of prime Westphalia, they have no objection to put up with good Yorkshire. Then their liking for cheese is so well known that it requires but few remarks to substantiate it. As regards this article, the stomachs of rats are of a most accommodating character. They are not particular whether it be Stilton ;—good old Cheshire, or Gloucester, single or double, all are equally welcome ; or, in the absence of more fancy cheeses, they have no objections to finish off their meal with a piece of dumpling Dutch.

I have been informed by cheesemongers, that rats will frequently drill holes through the flooring beneath large Cheshire or other cheeses, and then eat their way into them, and thus they will frequently consume pounds and pounds in a night or two ; nor is it an uncommon occurrence, they tell me, to find a large cheese with the inside scooped entirely out, leaving the rind a mere empty worthless shell.

I have heard a curious instance of seven rats working a Cheshire cheese-trap for their own destruction. It has been remarked by an eminent physician, that great eaters dig their graves with their teeth ! How far this may be correct, I will leave others to decide ; but, certain it was, that the rats I speak of dug theirs with their teeth. Four or five large Cheshire cheeses were placed on each other in a store-room, and the rats beneath the flooring drilled a hole through the boards, and so worked and ate their way into the bottom cheese. One night, however, it appears that, while they were busily engaged, the walls of the cheese, which had been rendered so weak and thin by the rats within, gave way, and down came the pile upon them. On the following morning, when the men removed the pile, they found the bodies of seven rats, all of which had been regaling themselves in the bottom cheese, when it fell in and crushed them in the ruins.

It appears from the accounts of several authors, that the rats of England, like the rats of Jamaica, have at times a liking for fruits. A gentleman residing at Battle, states that the renowned Battle Abbey abuts on his neighbour's garden, where he has a fine Morella cherry-tree growing against the wall. For some days past he had missed great numbers of cherries from his tree, and could not detect the thief. But at length, walking one day at noon, he saw a house-rat deliberately biting the cherries off by the stalks, and taking them away to a hole in the wall. This is a most extraordinary occurrence, as, the cherries being bitter, one would have imagined that they would be far from palatable.

I have been informed by several butchers in Newgate market, that rats are the most troublesome creatures they have to contend with; that every day in their lives all the meat they have left on hand, must be hung up out of their reach, or, in the night, they will so mangle and drill it, that nothing after can be done with it except for sausage-meat. It matters little whether it be mutton, beef, pork, veal, or lamb; to them all are the same. Nor is it always safe from them even when hung up; for if there be any possibility of clambering or jumping up, they will have it; and though we are apt to view them in a most mean and contemptible light, nevertheless they are epicureans in their way, for, should a hind quarter of beef by any accident fall or be left upon the blocks or benches, they will not eat the leg or shin—oh, no! but will plunge vigorously into the softer parts of the rump or sirloin, and thus destroy the most valuable joints.

On the whole, it seems perfectly clear, that in a colony of rats, various portions of them possess different likings and dispositions, and classify themselves accordingly. Hence it is, that in the spring one party will betake themselves to the fields, a second to the hedges and ditches, a third to the water-side, and a fourth to the game-preserves; while a fifth will remain at home in the farmsteads to the great annoyance of the good dame and her daughters. But even in the same individual animal, its likings and dispositions will vary according to circumstances. In the autumn, however, nearly the whole tribe come to one opinion, and resolve to return to the barns and ricks for winter quarters.

CHAPTER VII.

TESTIMONIES OF MODERN WRITERS AND NATURALISTS AGAINST THE RAT.

I SHALL now proceed to give the published testimony of various modern writers and distinguished naturalists, on the devastating ravages of the rat among poultry, game, grain, &c.; and, at the same time, adduce a variety of serious and interesting facts.

The "Zoologist" contains a paper to the following effect:—

"Towards the spring, the common brown rat leaves our stacks and buildings, and lives the summer through in the open fields, and many are trapped by gamekeepers. They breed among the corn, and, like the rabbit, burrow a hole in the ground, into which they draw a few materials, and there deposit their young. They are very voracious at this period, and will attack young leverets and game of all kinds. In the autumn they will assemble together and return again, with all their young, to the barns and ricks. But some of them," it says, "will betake themselves entirely to the woods, poaching for their livelihood, and, from their wilder mode of life, soon alter in appearance. Their body grows longer and more weasel-like, their hair more shaggy, the hairs of their mouth longer, and altogether they assume a more ferocious and determined character. The feet of some he has noticed were red, and he believes that an ordinary observer would suppose them to be of a different species."

A gentleman informs me, that about forty years ago he resided in the neighbourhood of Halcom Hall, the proprietor of which kept six vermin-killers and five gamekeepers. At some distance behind the mansion, and close to the woodside, it was a yearly custom to erect a large faggot-stack for the use of the establishment, and my informant tells me, that he has both seen and helped to destroy the vermin when the stacks were taken down, and that they have destroyed between four and five hundred rats at one killing, all of which lived upon game, poultry, &c., in the immediate neighbourhood.

In the "Popular History of Mammalia," the author says that the rat is fond of the seeds of leguminous plants, fruits, and the melon tribe, which makes it a dangerous enemy to the gardener, and that it attacks poultry also.

The editor of the "Treasury of Natural History," after describing the animal appearance of the rat, states that, whenever it conveniently can, it makes its hole very near the edge of the water, where it chiefly resides during the summer months, feeding on small animals, fish, and grain. It also haunts the corn-fields, where it burrows and breeds. When winter approaches, it draws near some farm-house and works its way into the corn-ricks, where it consumes much, but wastes more. It also destroys rabbits, poultry, and all sorts of game, and scarcely any of the feebler animals can escape its rapacity.

The following interesting anecdote is quoted from the "Gleanings of Natural History":—

"The damage done by rats is not confined to barns, corn-ricks, &c.; but is much greater in corn-fields than is generally supposed. As they infest the fields in summer, I have no doubt but that they destroy a great quantity of game, and that much of the devastation which has been attributed to stoats, weasels, &c., has been committed by rats. When these vermin abound much in a farm-yard during winter, but little game will be found in its immediate neighbourhood in the ensuing autumn. Gentlemen, therefore, who are particular in the preservation of game, cannot pay too much attention to the cleanliness of the farm-yards of their tenants. It will be a great advantage to both parties."

The author of the "Book of the Farm" pronounces rats to be the most troublesome of all vermin, because they harbour in the steading. They not only make every place they frequent dirty, but disgustingly so. The mischief they do in cutting holes in boarded floors, in undermining stone pavements, gnawing harness, consuming and wasting every eatable thing, and killing hens and pigeons while sitting on their eggs, is very considerable. He says the old black rat is nearly extirpated ; and the fierce, dirtier, more mischievous and dangerous brown rat has taken its place.

The talented compiler of the "Natural History of British and Foreign Quadrupeds" says it is the most destructive of all sucking animals, devouring our meat, poultry, game, birds, fish, corn, and other articles of food. Nor does its mischief end there; for it gnaws our furniture, clothes, books, papers, &c.

The author of "British Quadrupeds" states that its astonishing fecundity, its omnivorous habits, the secrecy of its retreats, and the ingenious devices to which it has recourse, either to retain its existing place of abode, or to migrate to a more favourable situation, all contribute to keep up its almost overwhelming numbers. It digs with great facility and vigour, makes its way with rapidity beneath the floorings of our houses, between the stones and bricks of walls, and often excavating the foundations of a dwelling to a dangerous extent. There are many instances of their fatally undermining the most solid mason-work, or burrowing through dams which for ages served to confine the waters of rivers and canals. He also states, that after they have had their fill in the corn-fields they will carry off a large quantity, and deposit it in their runs.

The editor of the "Naturalists' Library" speaks of the rat in the following terms. "Its food consists of almost every kind of animal and vegetable substances eaten by other quadrupeds. In granaries and corn-yards it is extremely destructive, committing its depredations in the former by night, and in the latter feasting at leisure in the heart of the stacks, where it frequently produces its young, and whence it cannot be expelled until they are taken down, when the quantity of corn it has destroyed is sometimes found to be enormous. In houses it feeds on bread, potatoes, suet, tallow, flesh, fish, cheese, and butter; in short, almost everything that comes in its way, including leather and articles of clothing. In the poultry-yard it destroys the young chickens, and in game-preserves commits similar depredations. Instances of its mutilating children, and even of its attacking grown persons, are known. In the fields it devours great quantities of corn, beans, peas, and other articles of agricultural produce; and as it is extremely prolific, it often inflicts serious injuries. When provisions fall short it migrates, sometimes in large bodies, to a more favourable situation; and when settled in a place where its supply of food is ample, it rapidly increases to an astonishing extent."

The compiler of the "Rural Cyclopedia" states that it is very destructive to chickens, rabbits, young pigeons, ducks, and various other domestic animals. Eggs also are a favourite article of food with rats, and are sought with great eagerness; in fact, every thing that is eatable falls a prey tó their voracity, and can scarcely be secured from their persevering and audacious inroads. He then quotes an extract from the "Agricultural Journal," wherein a gentleman is complaining that rats of an extraordinary size and fierceness troop about his old house at night, with a clatter which a little imagination and the stillness of the hour magnify into charges of cavalry. He also states that his little pigs were torn from their mother and devoured, in despite of the formidable defence of the old sow. Nor was it possible to calculate anything like the quantity of grain consumed by them.

The amusing compiler of the "Natural History of the Highlands," states that nothing will keep out these animals when they have once established themselves in our houses. He says they gnaw through stone, lead, or almost anything; they may be extirpated for a time, but you suddenly find yourself invaded with a fresh army. Some old rats acquire such a carnivorous appetite, that old or young fowls and ducks, pigeons, rabbits, and poultry of all kinds, fall a prey to them. Adepts in climbing as well as in undermining, they get at every thing, dead or alive; they reach game, although carefully hung in a larder, by climbing the wall, and clinging to beam or rope until they get at it. They then devour and destroy all that can be reached. He has frequently known them in this manner to destroy a larder full of game in a single night. They seem to commence with the hind legs of hares, and eat downwards, hollowing the animal out, as it hangs up, until nothing but the skin remains. In the fields to which these rats betake themselves in the summer time, not only corn, but game and eggs of all kinds fall to their share. That they carry off hen's and even turkey's eggs to a considerable distance, the author says is a fact; but how they accomplish this feat he should very much like to know, as they do it without either breaking the shells or leaving a mark upon them. I shall explain this mystery to him in another place.

Enough, I think, has already been adduced to convince even the most sceptical as to the predatory nature of rats, with regard to fish, flesh, fowl, and vegetable substances. But I will wind up the evidences of our numerous authors by giving the practical testimony of our Salopian friend, as published in the "Sporting Magazine.

The "Salopian Gentleman" says that the very great damage which is done to farmhouses, outbuildings, and drains, as well as to the potatoes, carrots, and Swede turnips, both when in the hods and while growing, must be apparent to every one; but the blame is often most wrongfully laid to hares and rabbits. Thus the quantity of poultry destroyed (the eggs taken while hens, turkeys, and ducks are sitting away from the house—the young of every kind devoured as soon as they are hatched, and can follow the mother about the fields) is a constant source of loss and vexation to the farmer's wife and daughters, who generally have all the trouble of rearing the poultry, and only a share of the profit for themselves. To the small landholder or cottager, where there happens to be water near, it is worse; his chief hope being the pig, and next the flock of young geese, turkeys, &c. If such is the known and well-ascertained loss and destruction of poultry, what must be that which in the fine summer months is going on amongst winged-game of every kind? No doubt many, like myself, have constantly found the nests of partridges or pheasants and wild ducks destroyed, and the eggs missing or broken. Sometimes this is done by a stoat, weazel, magpie, or other vermin; but six times out of seven it is by the rats then living in the hedge-rows; and bits of the shell may be traced to their holes. This is particularly the case where wild ducks are preserved; and it is scarcely possible to protect the old duck when sitting close at hatching time, if near the edge of the water—the place most often selected for nesting. If the unhappy duck does get her tiny brood into the water, the pike assails them there; and when she seeks safety on the shore, the rats on land attack them. He has bred many wild ducks, and can well attest the damage done by the rats and pike; and if the old duck is lucky enough to get away with a brood of twelve or fourteen young ones, it very seldom happens she can rear more than half, or one-third, of those that were hatched. The rats are sure to have a share, as they do of the young pheasants and partridges. How often will one rat pay visits to a coop, under which are chickens, ducklings, or other young things, and, before it is detected, carry off and kill half the brood. What then must be the number killed by them when unwatched and unmolested in their savage occupation amongst the unsuspecting young pheasants and other birds, while wandering about in every direction?

CHAPTER VIII.

PREDATORY AND DESTRUCTIVE HABITS OF RATS.

In this chapter I shall commence with the slaughtering propensities of the rat, as regards singing-birds, rabbits, poultry, &c., and conclude with their consumption of horses. But in so doing let me remark, that there is scarcely a charge which can be recorded against rats, where there are not thousands upon thousands of individuals to be met with who can vouch for the probability of such charge by parallel cases, in which they themselves have been eye-witnesses. Nor can we speak in any company about rats, without there being individuals present who have some charge or charges to prefer against them. Before giving a few statistical facts, in substantiation of the evidence adduced by the various authors quoted, allow me to state that the *Mus decumanus*, or common Brown Rat, will, wherever it may be stationed, follow the bent of its nature; and though our Salopian friend may give his experience of the rats of Shropshire, and Mr. Newman his knowledge of the rats of Derbyshire, yet they may rest assured that the rats of Lincolnshire can equal the exploits of the rats of Shropshire or Derbyshire, for unqualified daring or rapacity; that the rats of Nottinghamshire can equal the rats of Lincolnshire; and the rats of Devonshire can vie with the rats of Nottinghamshire. Moreover, we may take it for granted, that the rats in every county throughout England, Ireland, Scotland, and Wales, and indeed every country and island in the known world, where British and French shipping have traded, can equal the rats of the forenamed counties,—all of which will, like bad men everywhere, commit their outrages anywhere, when time and opportunity offers.

For twenty years, a gentleman residing at Hammersmith, has been an amateur, or fancier, of song-birds, and from time to time, has supplied the "Gardeners' Chronicle" with many interesting anecdotes of their personal history. At the commencement, he built a most commodious aviary, and spared neither pains nor expense in bringing together everything which nature or art could supply, to render it a perfect ornithological palace. His collection of birds was admitted by all competent judges to be one of the best existing. He possessed a great variety of foreign birds, besides the choicest specimens of nearly every songster of our woods and forests. It used to be his pride and boast that he had more birds than there were days in the year,—his number amounting to 366; and for years, all was one unchequered scene of harmony, elegance, and happiness. But alas! one autumn saw a perfect annihilation of this scene of beauty. The heavy rains had so swollen the river Thames, that the rats on its banks were forced to retreat in all directions, and, unhappily an army of them most unceremoniously billeted themselves upon this gentleman. Scarcely had they arrived, when they bored sundry holes in the flooring of his aviary, which he as punctually stopped by nailing plates of zinc over them. Yet he never supposed, for a moment, that the rats came after his birds. He had recourse to poisoned food and Harrison's pills to get rid of them; but all to no purpose, for they carefully avoided touching any of it. Thus things went on, till one morning he discovered, to his great horror, that they had made another hole, and carried off all his birds but eleven. These cruel butchers had clambered up the poles, and taken them from their perches while sleeping. The survivors he put into small cages; thus making good the old adage, of locking-up the stable when the horse is gone.

The grand object, however, which I have in view, is that of taking effectual steps while the horse is safe. Suffice it to say, that after the murder of 355 of his poor little innocents, he then, by dint of barytes reduced to a powder, phosphorus, red-herrings and sprats, commenced a war of extermination against these sanguinary vermin, and succeeded in annihilating them.

In the "Gardeners' Chronicle" a writer states that he has, for a long time, been a great rabbit-fancier; that he has an excellent outhouse fitted up for them, with an exercise-ground, and everything that can contribute to their health and comfort; yet in despite of all this, though he has had his hutches mounted on trestles with projecting ends, and traps laid in all directions to catch the rats, he has been able to raise only five young ones; although he has eight does and two bucks. This is lamentable; for with such a stock and convenience he ought to have raised as many scores, and would have done so, but for the rats. They will at all times kill and devour young rabbits. This is a fact to which nearly every rabbit-fancier in London (a very numerous body) can bear practical testimony. Rats are the principal enemies they have to guard against. He says, so determined were they to effect the downfal of his establishment, that they actually bored one of their holes through the bottom of one of his hutches, which was placed upon trestles, fifteen inches from the ground, and so situated that no rat of mortal growth could possibly get between the bars. If this says nothing for their honesty, it says a great deal for their perseverance, ingenuity, and powers of standing upon nothing. But I think there is little question about their standing upon each other's backs to reach anything; for I have seen them myself standing in a pyramid, five or six deep, when huddled up in a corner; and I have also seen several suffocated and pressed to death by those above them. This is no uncommon occurrence in large rat-matches, when from fifty to two hundred rats will be put in the pit at one time. But for rats to enter a hutch, and take the young ones away, in despite of the mother, is an amount of daring and courage rarely to be met with in smaller animals, except of the weasel kind; for a doe rabbit is both a desperate biter and fighter in defence of her young; and five or six times I have seen cats most cruelly bitten and beaten by doe rabbits.

I shall now give my own sad experience of the voracious habits of rats among rabbits, pigeons, &c.

In my boyish days my father carried on an extensive business in London. The house was a corner house, and had areas fronting both streets, which led into eight large arched cellars, that ran beneath the roads. My father was not a wine-bibber, consequently the cellars were not cumbered with many pipes of wine, either in glass or wood; but being indeed an honest Yorkshireman, his palate ran more in favour of a glass of good home-brewed ale. Suffice it to say, that only three of the cellars had lumber and coals in them; the other five, with the exception of a roosting-box for my pigeons, were empty. Now, what was to be done with these five cellars? thought I. Immediately my youthful imagination conceived the idea of turning them into a rabbit-warren. The idea was no sooner formed than I sat myself down to calculate the means whereby so desirable an object could be brought about. First, I must have my father's sanction to keep rabbits; secondly, how were these rabbits to be obtained, since nearly all my money had been expended in pigeons. Then came the number that would be necessary to stock the warren. I made a moderate calculation, and concluded that five does and one buck would be sufficient. As to wild rabbits, they were out of the question; so tame ones must be substituted in their stead. By this time my very heart and soul seemed centred upon a rabbit-warren. I shall never forget it! So to work I went, with all the zeal and determination of an enthusiast, to conquer every obstacle that might stand between me and the object of my ambition. Nor do I believe there is any period of a man's life, from the cradle to the grave, wherein his faculties are brought into more vivid action than in his youthful days, when he is striving to compass some contemplated hobby. My small capital was collected together, and every available source sounded to increase the sum. My credit was pretty good; nevertheless, when all the moneys were put together they fell infinitely short of the sum required. What was to be done? Why, there was no alternative but falling back upon the oldest banker I had, namely, my father. Indeed,

I may say this was the first bold enterprise of my life. Father was a shrewd arguer and a stern-countenanced man; therefore no liberties would do; but at the same time he possessed feelings as kind and sensitive as Charity itself. I lost no time in making myself acquainted with all the ramifications of a warrener's business, and ascertained all the best markets, as well as the wholesale and retail prices of everything in connection with my new project. Thus far I considered myself equipped for the struggle; but when should it take place? that was the question! I recollected once having heard my father say, if you have a favour to ask of a man, let it be after dinner, for then he is so composed and comfortable that he will grant you anything; but if you ask him before dinner you might as well ask a wolf for its fangs. I therefore resolved upon taking his advice, and come down upon him after dinner. This I did; and never did a hawk more keenly eye every movement of his intended victim than I did every muscle and feature of my father's countenance. With an apparent indifference I opened the budget by asking the following questions: first, what was his opinion of rabbits? if he did not think that they were very pretty, healthy, and profitable creatures? secondly, if he did not think that country gentlemen were very much to blame for not improving the breed of them in their warrens? He turned and fixed his eyes upon me with an expression of wonder, almost amounting to bewilderment; and then resumed his work without giving me an answer. Afterwards my father, who either was, or appeared to be, more than usually absorbed in his business, without looking at me, asked me what I had been speaking about, when I renewed the questions, as to whether he did not think that rabbits were very pretty, healthy, and profitable creatures; and whether he did not think also that country gentlemen were very much to blame for not trying to improve the breed of them in their warrens? After another brief silence and a pinch of snuff, both of which were resorted to for time to gather his replies, he thus expressed himself. As to rabbits, he thought them very pretty things where there was convenience for keeping them; as to the healthiness of them in a house, he was not prepared to give an opinion. With regard to the profit of them, that required some calculation and experience; and as to the improvement of the wild breed, that was a matter he should like to hear discussed, since there was no question but that it would be most desirable, if practicable.

This latter remark was just the thing I wanted. To work I went, by first assuring him that I knew by experience, that if a litter of young tame rabbits be allowed to run at large without molestation, they will soon become wild and hardy, and burrow and run as fast as the best of the common wild breed; and if he would only buy me five does and one buck, and give me the range of the cellars, I could, in six months, prove the thing to a demonstration; and, at the end of three years, supply half the warrens in England with large wild-stock rabbits. He now clearly saw the object I was aiming at, which was to have half a dozen rabbits for my fancy, and the cellars to keep them in. My father then set to work to raise up one objection after the other, which I as quickly cramped or reasoned away with all the earnestness of a poor barrister whose scanty means will not allow him to lose his case. Then came the most weighty of all his objections—the amount of money it would cost to keep them. Upon this subject I was armed to the very teeth with calculations, in black and white, and from which there was no flinching. I ran through the prices of all the various foods, the quantity it would take each week to keep them, and so on, for six months; then the average number of young they would have, and the expense it would cost to keep them till they were three months old, and then the price they would fetch at the poulterers; proving thereby the immense interest that must of necessity accrue from so profitable a speculation, all of which would come into his own pocket, since I did not desire one farthing for myself. I then finished by observing that, in addition to all these vast profits, at the end of eighteen months I could supply the table with fat rabbits for the whole family, from two to three days a week, the whole year round.

After a hearty and good-humoured laugh, he said, "Well, well, Jem, my boy; I'll see what I can do." That was enough—the victory was mine,—for his bare word was at all times worth more than two-thirds of the promissory-notes in circulation.

Suffice it to say, the morrow saw me master of six of the finest rabbits that money could purchase; and for twelve months they absorbed my every exertion and attention. Time went on merrily! Litter after litter came forth in rapid succession, and grew and strengthened, and, indeed, everything was thriving and prosperous. It was a pretty sight. The passers-by would throng round the areas to see them fed. It was only necessary for me to give the well-known whistle, and out came rabbits and pigeons in all directions, till the place seemed to be one mass of living creatures; and such was the confusion, that there was no possibility of counting them. There were all sorts, colours, and sizes. But to touch them, though they appeared so tame, they would dash and plunge like mad things, because no human hand had ever before been on them, and they would be shy and distant for days. They had burrows in the ground like wild rabits. Thus they prospered for about eighteen months, when, at a rough calculation, there must have been at least from a hundred to a hundred and fifty live rabbits. But on going into the cellars one morning, there lay the remains of four young ones. Their bodies and bones were entirely eaten away, leaving the heads and legs of two hanging merely by bits of skin, while the other two were without heads, leaving only a portion of their skins, feet, and ears. I concluded it must have been the cats. So to make all things secure, when evening came I fastened all the doors, and went to bed perfectly satisfied as to the safety of my rabbits; but what was my surprise and mortification, on the following morning, to find the contrary; for there lay the remains of twelve mutilated young ones, of various ages, besides a magnificent full-grown, fawn-coloured buck. The latter had only just been killed; for he was perfectly limp, and the warm blood was still oozing from a wound in the neck. In deep lamentation, I took it to a rabbit fancier, who, upon asking me a few questions, very soon explained the mystery. "Ah! my young friend," he said, "the rats, at last, have paid you a visit, and now its good-by to your rabbit-warren." And so it turned out, for morning after morning I threw out the ghastly remains of numerous dead rabbits; when at the expiration of about three weeks there were only five ill-looking old ones remaining, and they were so ragged and bitten, that they were disposed of immediately. It turned out that the rabbits, in burrowing, had got so near to the drain or sewer, that the rats smelt them, and soon set to work and broke through, and hence the entire destruction of my warren.

Still I had a consolation left, and that was my pigeons. They of course must be safe, because their house was nailed as high up against the wall as the place would admit of, and I had been advised also to get a guinea-pig, as it would have the effect of frightening the rats away. This I did, and paid a shilling for a very fine one. But alas! poor guinea-pig and pigeons! On entering the cellar the following morning, there they were, all dead. The rats had clambered up some loose boards that were placed against the wall, and thence to the pigeon-house; and there were the mangled remains of the tenants lying about in all directions, while some were entirely devoured except the feathers. But what grieved me more than all was the death of a beautiful dragon-hen, that was sitting on four eggs, two of which were her own, and the other two a present from a gentleman; and to me of very great value, because they were the produce of a valuable pair of almond tumblers, for which the gentleman paid five and twenty guineas. But on the very morning that these eggs were to have been hatched, the rats ate them up; and as for the poor guinea-pig, not seeing him, I imagined he was chasing the rats through the drains and sewers; but, on moving the boards that had formed a ladder for the murderers, there lay poor master guinea-pig without his head! Here then was an end to my live-stock.

Now I will leave the reader to judge how far I am indebted to rats. The loss of my live-stock in early life fully convinced me of the rapacious and omnivorous character of the rat. From that day to the present my eyes have ever been upon them; and everything I have read, seen, or heard in the interim (a period of thirty years), has convinced me more and more as to the bane they are to individual and national prosperity. Consequently, after a fair and dispassionate consideration of the subject, I came to the determination and resolve, from feelings of the most patriotic and philanthropic character, to grapple with the evil, and make it manifest to the world by facts of the most stubborn and unflinching verity.

As a gentleman was walking one evening late, two or three hundred yards from his own home, he heard a great noise in a hen-roost. A poor fowl was squealing in great distress; and he distinctly heard the gibbering and half-squeaking noise of rats. Having been a great sufferer by them himself, he soon concluded what was the matter. On the following morning he told his neighbour of the noise he had heard over night in his hen-roost: "Ah," he said, "one of our old hens was killed last night, and gnawed all to pieces."

A few years back I took a trip to Hanwell, for the purpose of gathering some practical information from a celebrated ratcatcher, of the same place. Among other information, he told me of a gentleman, residing at Greenford, who sent for him the week before, after having had two fine hens killed by rats while sitting, the one upon fifteen eggs, and the other upon eighteen. I mention the number of eggs, in order to show the large size of the hens. They were gnawed all to pieces, and every one of the eggs taken away. The ratcatcher, by dint of ferrets and traps, succeeded in catching eighteen of the delinquents, for which the gentleman paid him most liberally.

A gentleman living at Hanwell, also sent for him the same week, after having had two full-grown ducks killed by the rats. Here there was ocular demonstration of the fact; for the gardener, early in the morning, saw two large rats in the act of dragging one of the ducks away, when he pursued them, and succeeded in getting it from them. The duck was then warm; but the head was gnawed all to pieces.

In the "Gardeners' Chronicle," a gentleman of Hampshire writes to state that in his district the neighbours have of late lost many hundred heads of poultry by rats. They will come in the night, and take the chickens from under their mother's wings. Some of them have lost as many as ten in a night. Nor are their depredations carried on in the night only; for they will wait and rush out at the chickens in the daytime, and carry them into their holes.

This gentleman has seen a hen pursuing a rat that was running away with one of her chickens. Last week some of the neighbours waged a war, and with the aid of dogs and ferrets they succeeded in killing above two hundred of these robbers.

A baker residing in London, had a spare kitchen, wherein he kept a number of fowls and four ducks; and from their being allowed to run into the bakehouse, which was kept continually warm by the oven, they throve pretty well. One of the ducks was sitting upon a dozen eggs; and, on the morning of hatching, curiosity prompted him to see how they were getting on. He found seven hatched, and the other five were in progress; for the shells of all were broken, and the bills of the young ones peeping out. In about an hour afterwards, as the family were above-stairs, they thought they heard the old duck making a very strange noise, but concluded that the whole of the eggs were hatched, and that the mother was making a great fuss over her young family. Curiosity of course led them to go gently down stairs, and have a peep at them, when, to their utter astonishment, there were two old rats standing upon their hind-legs, with their ears thrown back, their mouths wide open, and making an anxious moaning noise close by the old duck, which was backed into a corner, and pecking at them with all her might, or rather striking them with her bill. The clandestine old thieves were creeping nearer and nearer, evidently with the intent of fixing upon her as soon as an opportunity should offer. So intent were they upon their savage design, that they did not perceive the person approach them; whereon he dealt one of the old savages such an unmerciful kick as sent him flying, and the other was off before he could kick him. But on looking after the young ducks they were all gone, eggs and all. The rats had carried them away into their holes; and these two old cannibals were trying what they could do with the mother; and doubtless, had they not been so unceremoniously disturbed, they would have succeeded in killing her also.

Lately a gentleman, of whom I have already spoken, was greatly annoyed by the following circumstance. He had eighteen fine young goslings, more than a week old. They were all hearty and strong, and bid fair to make an early and profitable flock of geese. One evening he penned them safely up with the old ones, and having several large dogs about, they were considered perfectly safe from thieves. But when he opened the pen in the morning he found, to his unutterable surprise, that there were only five young ones. He could scarce believe his own eyes; but on recovering himself a little, and making a casual survey, the mystery was soon dispelled, by discovering an aperture which the rats had made in the night. They had entered the pen, and in despite of the old geese had succeeded in destroying thirteen goslings out of the eighteen, carrying twelve away, leaving one dead and five living behind them. And it is believed that had they not been disturbed by the domestics rising early to their occupations, they would have succeeded in carrying away the whole eighteen.

This gentleman has a great number of ducks, and a very valuable collection of Malay fowls, over which he is very choice; so much so that he declines letting any one have their eggs for sitting. These fowls are very valuable; for in London, I have often been asked from ten shillings to two guineas apiece for them, and at the moment I am writing I should be most happy to give any one ten shillings for ten pure Malay eggs for sitting. But this gentleman informed me that one summer the rats robbed him of above two hundred of these valuable chickens, besides a host of ducklings. Now the losses he has sustained by rats are truly alarming. He has plenty of fine meadow-land for the fowls to run and feed upon, and the great question is, whether these chickens which the rats destroyed would not have more than paid his rent. I believe they would, taxes and all. Only think of this; it is worth the gravest consideration.

In concluding this portion of my subject, I shall proceed to notice the omnivorous character of the rat; for I believe there are facts enough already quoted to satisfy even the most scrupulous inquirer as to their predatory and destructive habits among poultry and game.

Mr. Jesse, in speaking of the Cape geese, which are kept in the large ponds in Richmond Park, says they used to build their nests on the island in one of those ponds. In consequence, however, of their eggs having been so frequently destroyed by the rats, the geese took to building their nests in some oak pollards near the water, from whence they conveyed their young in safety. But how they got the young down is quite uncertain, the keeper believing they took them under their wings, and then descended the tree; but Mr. Jesse believes they bring them down as a cat would her kittens, namely, in their mouths. He also knew a poor duck in Staffordshire, which built its nest in a poplar-tree by the water side, to get out of reach of the same enemy.

The increase of rats (says Mr. Jesse), where little pains are taken to destroy them, is something enormous. The females are said to breed five and six times in the year; and they will produce twelve, fourteen, sixteen, and even as many as eighteen at a litter. The most interesting account of rats he ever met with was made some time ago in an official report to the French Government. It was drawn up in consequence of a proposition that was made for the removal of the horse-slaughterhouses at Montfaucon to a greater distance from Paris; when one of the chief obstacles urged against such a removal, was the fear entertained by the inhabitants in the neighbourhood as to the consequences that might result from suddenly depriving these voracious vermin of their accustomed sustenance. The report goes on to state that the carcasses of horses killed in the course of a day (and sometimes these have amounted to thirty-five) are found the next morning picked bare to the bones. Monsieur Dussaussois, a proprietor of one of these slaughterhouses, has, however, made a still more conclusive experiment. A part of his establishment is enclosed by solid walls, at the foot of which are several holes made for the rats to run in and out. Into this enclosure he put the carcasses of two or three horses, and towards the middle of the night, having first cautiously, and with as little noise as possible, stopped up all the holes, he got together several of his workmen, each having a torch in one hand and a staff in the other; then, having entered the yard and closed the doors behind them, they commenced a general massacre. It was not necessary to take any aim; for no matter how the blows were dealt, they were sure to kill a rat; and those which endeavoured to escape by climbing up the walls were quickly knocked down. By a repetition of this experiment at intervals of a few days, he killed, in the space of one month, 16,050. After one night's killing, the dead amounted to 2,650 ; and the result of four hunts was 9,101.

Even this can give but an imperfect idea of the vast number of these vermin ; for the enclosure in which they were thus killed contains not above one-quarter of the space over which the dead bodies of horses are spread, and which it is but fair to suppose must equally attract the rats upon all points. These animals have made burrows for themselves like rabbits in the adjoining fields, and hollowed out into catacombs all the surrounding eminences; and this to such an extent, that it is not unusual to see them crumbled away at the base, and leaving their subterraneous works exposed. So great is the number of these animals, that they have not all been able to lodge themselves in the immediate vicinity of the slaughterhouses ; for paths may be distinctly traced leading across the fields, from the enclosure in which the horses are killed to a burrow about five hundred paces distant. These paths are particularly remarkable in wet weather, being covered with a clayey mud, which adheres to the feet of the rats on running out of their burrows.

The liking which these animals show for one particular part of a horse is curious. They invariably begin by devouring the eyes, and eating the fat at the bottom of the orbit. There has not been one instance of a dead horse being left for one night exposed when the eyes have not been devoured before morning.

During very severe frosts (when it becomes impossible to skin and cut up the bodies of horses that have been exposed to the air, and when even the fragments of flesh lying about have become so hard as to render it difficult for rats to feed upon them), they eat their way into the bodies of the horses, and there establishing themselves, devour all the flesh ; so that when the thaw comes, the workmen find nothing but a skin and skeleton underneath, as clean and clear of flesh as if it had been prepared by the most skilful operator.

The conclusion of the French commissioners' report (says Mr. Jesse) is of too great importance to be omitted. It contains a useful hint to those who may be inclined to frequent the minor restaurateurs, or humbler eating-houses of Paris. It goes on to state, that a man and a woman are employed the whole year round in flaying and dressing up dogs and cats.

The skins of the dogs are hung up to dry, and those of the cats are carefully stuffed with straw; the fat of both is carefully collected and melted down, and the paws and tails are sold to the glue-makers. The commissioners say they never entered this establishment without finding a great number of dogs and cats flayed, disembowelled, and trussed with the greatest care, and quite ready for the spit or stewing-pan. The heads and tails are always cut off; these dogs and cats, thus prepared, present a very tempting appearance, and it would be difficult to distinguish them from animals of the same size which are admitted to our tables.

Here let me remark, that it is a pity that the gourmands of Paris should look down upon these humbler eating-houses with such an air of contempt; for it must be a source of gratification for them to know that they are never likely to experience any want of frogs' legs to make patties and fricassees, while there are plenty of young rats at Montfaucon.

CHAPTER IX.

THIEVISH PROPENSITIES OF RATS.

THE furtive habits of rats are among the most alarming evils of domestic life, and often pregnant with mischievous consequences in the family circle. It is not unreasonable to suppose (indeed I have known such to be the case) that domestics have from time to time been suspected of frauds and thefts, and discharged without characters; and neighbours and acquaintances have struggled under gross suspicions of crimes of which they have been perfectly innocent, while the midnight marauders and furtive rats have been the real delinquents.

I recollect reading in the papers some years since a very distressing circumstance of this kind. The sufferer was an interesting orphan girl, the daughter of once respectable parents, but who had been, by the death of her parents, reduced to the capacity of an ordinary servant. Her master and mistress, who were somewhat advanced in years, were tolerably kind, perhaps more so on account of the girl's misfortunes; but one morning it happened that two very valuable gold rings belonging to her mistress were missing. What could have become of them? Who had seen them? Her master had not; and her mistress was certain that she had placed them on the table the night before. Then what could have become of them? Sarah did not know; master did not know; nor did mistress know; but yet two things were certain: first, that no one whatever had access to the house; and, secondly, that no one had been in the room but Sarah. The consequence was, poor Sarah was urged to make a special confession of her crime, and bring the rings back, in order that no further proceedings might be taken in the matter; but Sarah, conscious of her innocence, declared that she knew nothing of them, and, with all the dignity of wounded maiden honour, gently reproved them for their ungenerous suspicions. This, of course, was construed into hardened guilt. The result was, she was exposed before some of their friends, and in charity was not prosecuted, but sent about her business without a character, to the great grief of poor Sarah's friends and neighbours, who believed her perfectly incapable of any such offence.

About six months afterwards there was a terrible stench arising from beneath the flooring in one of the rooms. A carpenter was sent for, who soon took up the boards. Nothing was to be seen that could cause such a stench; but, on raising the hearth-stone, there were the putrid remains of a large dead rat. And what else? why, the remains of a rat's-nest! and what was in it? the two identical rings that were lost. The thing was proof beyond all doubt. The master and mistress were both of them satisfied of the wrongs they had done poor Sarah, who was instantly sought for and found, and honourable reparation made by her master and mistress, who, having no children, adopted her as their own, and left her sole heiress of all their property.

Some individuals may still have their doubts as to how the rings got into the rat's-nest, and probably suspect poor Sarah's innocence. Such persons can, perhaps, explain the following curious circumstances, which I found in an Edinburgh paper. It states that, while some men belonging to

the old gas company were putting a gas branch into a shop, they fell upon a rat's-nest beneath the pavement, containing one yard of pack-sheet, and a letter. The letter was dated a month back, with the seal unbroken, and the gas-men were putting the pipe into the shop of the very individual to whom the letter was addressed. How the letter or the pack-sheeting found their way into the rat's-nest is a mystery.

One night, the overseer of a farm, laid his watch on a table in his sleeping apartment, previous to going to bed. Towards morning he was awoke by a crash of something that had fallen, and a rattling sound, as of something being dragged along the floor. He immediately got up, and found his watch was gone. He lost no time in pursuing the thief, following the direction of the sound, when he came upon the watch at the mouth of a rat's-hole, into which the rat had entered, taking with him the whole of the guard-chain, and was only prevented from taking the watch by the case springing open from the fall, which made it require more room than the hole would admit. As it was, the rat did not seem disposed to lose his prize, but kept a firm hold of the guard, when the owner tried to pull it from him.

Nature has her own way of accommodating herself to some of the most extraordinary conditions of life and being. A gold wedding-ring was lost by the lady of the house, who for some reason removed it from her finger, one night after supper, and left it on the table. In the morning it was gone, and no one knew anything about it, which of course caused a great deal of misgiving and unhappiness. However, some three years after, as the master of the house came downstairs, the cat, with all the solicitude of one accustomed to be praised, came purring and mewing about his feet, and laid down close by them a large rat, from which she had eaten the head. The neck of the dead thing was exposed, and as her master stooped to caress her, his eye caught sight of some shining metallic substance, round the neck of the rat. It was a ring, completely embedded in the flesh. How could it get there? and what ring was it? were questions passing though his mind. By this time another inmate of the house came downstairs, who having his attention directed to the rat's neck, proceeded carefully and scientifically to

examine it, and soon pronounced it to be a small gold ring, which could not have passed over its head but at a period when the creature was very young. Upon closer examination it was found to be the lady's wedding-ring, lost so long before, which must have been carried away by the parent-rat to the nest of her young ones, and being dropped into that quiet home, the struggling blind creature must have thrust its tiny head through the golden noose. Each day the rat enlarged in size, making it more and more impossible to remove the ring, It was now a permanent collar, but of such small dimensions that the wonder is, how nature continued to permit her living demands to be supplied. Here were the two great canals of life to the brain, and other two canals bringing back the blood from the brain, the gullet, and the windpipe, all compressed into the narrow space of a wedding-ring; and yet the creature lived—was fat, and full of flesh. Nature must have been most accommodating in her efforts to sustain vitality in this little quadruped. She must have done a wonderful thing to permit life, health, and vigour to exist within the slender limits of a wedding-ring.

CHAPTER X.

THE DESTRUCTION AND EXTIRPATION OF RATS.

A FEW years ago, the journals teemed with an account of a grand rat-batteau, or rat slaughter, in the sewers of Paris; and that two merchants of Grenoble had contracted for the skins. In a few days, by dint of mutton-suet and leathern bags, they caught no less than 250,000 and at the expiration of a fortnight they had killed above 600,000. The merchants of Grenoble had agreed to give ten francs a hundred for the skins, which was at the rate of a penny each, and amounted to £2,500 English money. The merchants were obliged to renounce their contract, not having capital enough to complete it; and it was afterwards clearly ascertained, that a leather-dresser, in the city of London, bought them at an advanced price.

Now comes the grand subject for our consideration. In

the first place, can ten sewer rats eat a pound of beefsteaks for breakfast; that is to say, one ounce and a half each, for the eight smallest, and two ounces each for the two largest? No one, I think, will doubt it for a moment. Well, then, we will set down ten rats to eat a pound of beefsteaks for breakfast. In the second place, we will calculate a fair bullock to weigh half a ton; I mean, without the bones, hoofs, horns, and hide; that is to say, 1,120 pounds of eatable substance.

Now then 600,000 rats, at ten rats per pound, will eat 60,000 pounds of beefsteaks for breakfast; or, what will perhaps bring it better within the range of our comprehension is, that it will require fifty-three bullocks, at half a ton each; and besides these, there must be 640 pounds more cut off the fifty-fourth, to furnish them with a breakfast. This, I think, will tend most materially to remove all doubts with regard to the thirty-five horses at Montfaucon.

Here, perhaps, prejudice will cry out, "Down with all the rats in the universe, and annihilate them!" But reason says, "Stop—no haste—because there are two sides to the question! In the first place, there is no doubt but that 600,000 rats would consume more than an equivalent to fifty-three bullocks and a half for breakfast; and, in the second place, there is no question but that they would eat the same amount for supper. Then, on the other hand, it is not more than reasonable to suppose that they did not catch one half of the rats in the sewers; but, for reason's sake, we will set them down at one-half. That would make altogether, with those that were caught and those that escaped, 1,200,000 rats, which would eat, in animal and vegetable matter, an equivalent in weight to 214 bullocks every four-and-twenty hours.

Now let us multiply these by seven, as a week's sustenance for the rats in the sewers of Paris, and we shall find it amount to 1,498 bullocks,—or, what is equal to it, 749 tons of animal and vegetable matter.

Let us now suppose the rats to be removed for two months—say June and July—and what then would be the consequences? Why, that there would be an equivalent to 11,984 bullocks, or 5,992 tons of animal and vegetable matter, rotting and putrefying in the drains and sewers of Paris.

Now let us make one more calculation, by supposing the rats to be removed for one twelvemonth; and then let us see what would be the result. It would be thus:—There would be an equivalent, in weight, to 77,896 bullocks, or 38,948 tons of animal and vegetable matter rotting in all the various stages down to putrefactive and destructive fermentation.

What, then, would be the state of the drains and sewers of Paris—nay, of the river Seine itself? Would not every hole and corner in the capital be thoroughly charged with the noxious vapours, and a deadly pestilence hang over the city, like the angel of death shaking down destruction on all beneath; so that at the end of eighteen months there should not be one human being left to tell the fate of all the rest?

Here I will leave the matter for the consideration of those whom it most concerns, and conclude by warning the authorities of Paris on the probable consequences of killing their sewer-rats by 600,000 at a time!

I shall probably be asked by citizens on every side, if I hold out no hopes of relief from the ravages and devastations of this destructive and all-devouring animal? To such inquirers I answer, most emphatically, "Yes!" I give them every hope—nay, more than hope; for if they follow strictly the rules which I here lay down, they will see the matter reduced to a positive certainty, and find themselves as free—nay, more free—from rats than thieves of almost any other denomination.

I shall now lay down unerring plans for getting rid of all the rats that infest our houses, warehouses, &c. In the first place, send for an honest bricklayer; not a ratcatcher, for the latter, in London, are just about as scarce as stray diamonds.

I might as well here put you on your guard against a class of individuals who, of all vagabonds that prowl the streets of London, are the veriest scoundrels. They carry on their private callings under the outward guise of ratcatchers, or dog-dealers, or more properly speaking, *dog-stealers*. They are in connection with many of the thieves and housebreakers of London, and who, of all individuals, merit the vigilant and scrutinizing eye of those most valuable of all public functionaries—the police. Their private avocations are thieving, dog-stealing, and pioneering for burglars. When you once call one of these individuals in for the purpose of destroying your rats, he claims the privilege of prowling at will through your entire establishment, under the pretence of ascertaining the holes and runs of the rats. Thus does he ransack every hole and corner, from the garret to the cellar; and here and there, where he deems it convenient, he will deposit some cake, oatmeal, or biscuit, and with it a dead rat; but whether he uses poison or not is a matter of no consequence, since no one ever thinks of testing it by tasting it. Thus does he practise his deception upon the unwary, and will work for anything they choose to give him. Indeed you can engage one of these worthy functionaries for anything between pence and pounds; or they will sell you any amount of their poisoned cake, from a penny-worth to a crown's-worth. They are by no means nice to a shade, as sixpenny-worth of coarse oatmeal and a penny-worth of sugar, with a little grease and a frying-pan, will make a pound's-worth of their secret rat poison. They are perfectly indifferent as to pay, since they do not depend upon you for subsistence, but look sharply out after the gold and silver rats that chance may throw in their way in the shape of spoons, watches, &c. Nor are they by any means particular as to the shape or weight, but at all times prefer gold to silver. By-and-by the dead rats are brought forward as a proof of their professional skill. But after you have paid him well for poisoning a few dead rats, and he has fairly decamped from your premises, just have the kindness to see if all the door-keys, bolts, and shutter-fastenings are safe, or you may be visited in the night by another class of ratcatchers, who are made perfectly acquainted with all the nightly locations of every individual in your house; and if you have a house-dog, before going to bed just see if he is not lying prostrate, and, from his convulsive appearance, showing every symptom of death before morning.

Nevertheless, in a large place like London, there are some practical ratcatchers to be met with, who, in their way, are as honest as most other people. But these men are mostly responsible individuals, being house and shopkeepers, consequently seldom prowl the streets for sixpenny, shilling, and eighteenpenny jobs. Their time is of too much value; but, in their professional calling, they mostly *contract* with large establishments, at so much per year, regulated according to the number of times they are expected to visit the premises during that period, and on conditions also that they have all the rats they catch for themselves, which are afterwards sold in their shops at sixpence each, for training dogs and rat-matches; but in most cases sewer-rats are forbidden in rat-matches, as they are too large and foul-mouthed to risk valuable dogs with.

Here let me warn the reader, that should he at any time be bitten by a rat, let the wound be thoroughly cleansed immediately with warm water and yellow soap; but in all cases, when a valuable dog has been rat-killing, they bathe his wounds, to cleanse them from any venom that might be hanging about any of the rats' teeth.

A well-known ratcatcher reported to me, that he kills rats, both by the year and by the job. The proprietor of the Tavistock Hotel, Covent Garden, has employed him for years past, at six pounds per year. He visits the hotel weekly, and he informs me that he brings away from twenty to thirty rats each visit. Thus, if we divide the difference, and set them down at twenty-five per week, it amounts to 1,300 rats in the year, and these rats he sells at sixpence each, which produces £32. 10s. Then add his salary to this, and we find, according to his own statement, that he makes £38. 10s. a year by the rats of the Tavistock Hotel. Here then is a guarantee that this man is an honest ratcatcher; for, in the first place, he would not be trusted all over the hotel weekly, if time and experience had not proved him such; and, in the second place, were he otherwise disposed, his living is too valuable to risk for a paltry act of dishonesty; and there is no doubt but there are many other ratcatchers in London who may stand upon a perfect equality with him; therefore let it be distinctly understood, that I do not wish to say one word in depreciation of the true class of honest ratcatchers; but to the utmost of my power I will warn the public against that class of vagabonds, who, by putting on the guise of ratcatchers, as a cloak, bring an honest profession into disrepute.

The same man is also engaged at the Lambeth Workhouse

at six pounds per year; here he only pays a monthly visit, and is, according to contract, compelled to poison all; consequently he can form no idea as to the numbers killed. He likewise informs me that, some time ago, he was engaged by a gentleman of New Galway, to free his estate from rats. That gentleman defrayed all his expenses there and back, and gave him half a guinea a day, besides supporting him upon the fat of the land. It took him six weeks to complete the job, for which he received £18. 18s.

Having disposed of the ratcatchers for the present, and exposed the wolves in ratcatchers' clothing, we shall proceed to give some unerring methods for clearing one's premises of rats; at the same time I propose making every man, if he pleases, his own ratcatcher, and thereby entirely supersede burglars, both within and without.

In the first place, as I said before, send for an honest bricklayer, and let him hunt out every rat-hole in the foundation of your premises; and into each, where practicable, put a quantity of unslaked lime. After this, let him plaster every hole tight up with broken glass and mortar; then you may rest assured that the first time rats come that way, they will run back much quicker than they came; for if there be one thing in the world that rats hate more than another, it appears to be quicklime; consequently, if a place be strewn with it, or if some be placed round about their holes, the instant they find it out, they will beat a retreat, and not come that way again till it is gone.

In proof of the efficacy of lime, I shall instance a gentleman who had occasion to use a quantity about his premises, which had hitherto been much infested with rats, when, before using the lime, he could scarcely walk across the yard at night without treading on them. But he placed some fresh unslaked lime around their holes, and it had the effect of driving them all from his premises.

As an extraordinary instance of rat-killing, I may mention that a gentleman living in Wales, aided by his two juvenile brothers, killed no less than 170 rats in less than half an hour, by the use of lime. The three brothers poured some water into some quicklime, and after stirring it, poured it into the rat-holes, when out dashed the rats, heels over head, one over the other, and were as quickly killed by sticks, which the youths held in their hands. No dogs whatever were used upon the occasion; and the dead rats filled a bushel basket.

With a full knowledge of the value of quicklime, let the bricklayer secure all the holes, and put the drains in complete repair, and after that well line them with a thick layer of cement. But if they are very much dilapidated it will be much cheaper and much safer to have them done with pottery drain-pipe, with a trap at the end; for pottery-pipe is entirely proof against the teeth and claws of rats; while, on the other hand, they will gnaw through both lead and zinc. But should you have them repaired with brick, let him run iron gratings in grooves, at the ends; then, if at any time there should be a stoppage, it will only be necessary to draw up the gratings, and the obstruction is removed; but be sure to let them down directly after. By this means the rats will be kept most effectually out of your premises, and all the annoyances and expenses they otherwise incur be entirely obviated.

Here I may remark that, in London houses, rats are very much on the decrease, and principally on account of the extensive use of pottery-pipe for all purposes of drainage, as also the new contrivance, or trap attached to them, whereby the vermin are kept in the sewers.

Though the bricklayer may have most effectually fastened the rats outside your premises, still it is ten to one but he has as effectually fastened some inside. How then are these to be disposed of? I will tell you. In the first place, let a sharp look-out be given to ascertain their runs; that done, do not frighten or disturb them, but leave them alone; or if anything be done, throw them down some crumbs of bread and cheese, or anything else, and it will draw them that way again; for they will always go where they find food, and are undisturbed.

Now I will give you your choice of three modes of destruction;—the first is poison; the second is ordinary trapping; and the third, Uncle James's infallible rat-trap, that acts without baiting; all of which you will find under their respective heads towards the end of the book, and where also you will find the grand secrets of the profession, with the most approved methods of baiting, mixing, and using both traps and poisons.

But I have three strong objections to poisoning in dwelling-houses, especially where the end can be achieved without. The first is, the danger of accidents through such deadly articles lying about; next, the torture the poor creatures suffer before death; and lastly, the most offensive and disgusting nuisance they create by dying and putrefying beneath the flooring and behind the wainscoting of our apartments; so that often a carpenter is forced to be called in, and pull the place to pieces to dislodge them. But all this is entirely obviated by the use of Uncle James's infallible rat-traps.

I lay it down as a fundamental principle that one practical and demonstrative fact is worth ten thousand theoretical and speculative opinions; consequently, if I refer you to one easily ascertainable and well-authenticated truth, established upon seven years' practical experience, it will be worth all the floating fantasies of an age.

A cheesemonger in London (of whose losses in wearing-apparel I have already spoken), was in the habit of using his front kitchen as his store-room, wherein he kept cheese, flour, butter, hams, eggs, &c. The back kitchen was the bakehouse, and the oven ran into the yard. He carried on an extensive baking-business, and used to have a large stock of flour in at a time; for years he had been terribly infested with rats, which had caused him a deal of uneasiness; for daily and hourly there would be fresh proofs of their devastations, and the losses he sustained were enormous. Like thousands besides, he merely grumbled, yet never thought of resorting to some effectual means to put an end to the nuisance. He says the rats would make two or three holes in the sacks, and the floor every morning would be thickly strewn with flour, which, from their running over it with their dirty feet, was rendered perfectly useless; and some of the sacks would not have two bushels of flour left in them. He calculates that they used to eat and waste from half a bushel to a bushel of flour every night. Then when he went to fetch a cheese or two to place in the shop for sale, he would find some a quarter eaten away, and others half-gone. The butter also suffered most alarmingly;

and as for the hams and bacon, they were a luxury with them. Thus were his losses beyond all calculation. They used also to walk upstairs into every room in the house, and would take off any small article of wearing-apparel they took a fancy to.

At last, so numerous and endless were their depredations, that our victim seemed to have entirely resigned himself to his fate. One day, however, he took it into his head to have his warehouse altered, in order to make room for a larger stock of flour. Consequently he sent for his bricklayer, to knock down the projecting chimney, and also the copper flue and copper fixings,—indeed to clear everything away, so that there would be room for about twelve more sacks of flour. The bricklayer went into the warehouse to examine the old chimney, and on looking up to see how it was situated, he there saw, to his astonishment, that the inside of the chimney was completely lined with rats. He told me they were crammed in all the crevices ; and on the projecting bricks they were standing upon each other's backs two and three deep. This, it appears, upon the least alarm, was their retreat. The bricklayer walked quietly away into the area and cellar that ran beneath the street, to see if there were any holes to escape by. He only found one, and that was about the size of his hat, and led direct into the sewer. This he speedily stopped up, as well as the drain. Then there was a hole in the back of the chimney, which led into the coopers' kitchen in the next house. This he dared not touch for fear of disturbing the rats ; so he secured the door, and went for his dog Pincher, at the same time putting the coopers on the alert, who quickly took their stations with sticks and staves in their hands to knock them on the head as they jumped through. The bricklayer took his dog down, and secured the door, then, with a well-directed blow peculiar to the profession, he brought down some feet of the chimney at once. In an instant there was nothing but dust and rats flying in all directions. The dog set to work, and his master hammered away most lustily, while the coopers were roaring and banging with all their might. Thus the battle and confusion lasted for about ten minutes, when, for the want of more rats, something like tranquillity was restored. The battle now was over. The coopers killed about two dozen but on gathering the rats together, there were 139 ; and besides these, there was a nest in the flue with eleven young ones in it—thus making in all 150 dead rats.

Now what was to be done to prevent this house from being invaded by a fresh army of these destructive animals? Why he adopted the plan which I have already described with regard to drains, sliding-gratings, and rat-holes. The result has been that from that day to this, a period of above seven years, not a single rat has been seen or heard of within the premises.

Some years ago a correspondent of the "Boston Cultivator" recommended potash for the purpose of driving away rats. The rats troubled him so much, that he felt justified in resorting to extreme measures to effect their expulsion from his premises. He pounded up potash, and strewed it round their holes, and rubbed some under the boards and on the sides where they came through. The next night he heard a squealing among them, which he supposed was from the caustic nature of the potash that had got among their hair or on their bare feet. They disappeared, and for a long time he was exempt from any further annoyance.

CHAPTER XI.

WONLERFUL TALES OF RATS.

Before I treat upon the courage of rats, and their reasoning powers in attack and defence, I will first explain how they carry eggs; but not as the Fables of La Fontaine represent, which is by one rat getting over the egg and gathering it up between its four feet, and then rolling on its side and back with the egg upon its stomach; while a second rat takes him by the tail, and, throwing it over his shoulder, drags him backwards to his hole. All this is very pretty for a fable ; but still I have met with persons credulous enough to believe almost any exaggeration that may be ascribed to rats ; and if the two following accounts can meet with credence, then I must say that there are no rats in the wide world, except the American rats, that can equal the ingenious rats of Scotland.

An extraordinary instance of the sagacity of rats was witnessed in Argyleshire, a short time ago, by a man who was watching a quantity of eggs lying on the quay with a view to be packed. Four large rats were observed by him to issue from a hole, and rush all at once to the eggs, when presently they rolled one of the eggs to a distance from the heap, while one of the party, falling on his side, firmly embraced it with his legs; then being turned on his back, the rest yoked themselves to the burden, two in front and one behind, and by this living vehicle they safely conveyed their booty to the hole. The man was so amused with the contrivance, that he gave no alarm, nor caused the robbers any molestation.

An evening or two after, I was in the company of a gentleman who related the above story which he had read in the newspaper, and, feeling jocularly disposed, I made a few remarks upon the unenviable situation of the prostrate rat, as I supposed that the only way the two front rats could yoke themselves to the living vehicle would be by running their claws through his ears, while the hinder one would bury his in his haunches, for the purpose of raising him up, and forcing him along wheelbarrow fashion. These remarks aroused the Scotchman, who, suspecting I was ridiculing his countrymen, rose, with all the richness of his native diction, to put me down for my wanton incredulity, by relating what he called a fact; for he himself had been an eye-witness to it.

As he was walking one morning by the side of the railroad, a few miles from Glasgow, his attention was attracted by two rats which ran across the line to a hedge on the opposite side. They appeared to be male and female, as one was much larger than the other. He concealed himself for a time to watch their manœuvres, when he presently saw them merging from the hedge bottom, the one dragging the other by the tail. The female was on her back, with a hen's egg cuddled closely up between her four legs, thus making a cart of herself, while the male rat, with her tail over his shoulder, was dragging her along like a truck. All went on very well till they came to the iron rail, which stood about sixteen inches from the ground. How were they to pass over this? there was the difficulty. He presently let go her tail, seemingly to have some conversation with her as she lay on her back cuddling up the egg. After this he laid hold of his own tail, and placing the tip of it in her mouth, commenced clambering on to the rail, and as soon as he arrived at the top, he sat himself astride like a little jockey, and pulling up his mate, who held tight to his tail, then let her gently down on the other side, and in this manner they passed the four rails, one after the other, without hurting the egg. He then took her again by the tail, and throwing it on his shoulder dragged her backwards into their hole.

Here the narrator paused for approbation. The ladies stared with astonishment, and the gentlemen seemed perfectly dumbfounded at the surprising ingenuity of the "Rats of Scotland." But for my own part there was one thing I could not reconcile my mind to, and that was the difficulty of holding her by the tail with his paws, seeing they have neither fingers nor thumbs. I could perfectly understand rats hanging by their claws; but as to their grasping anything with their fore feet, sufficient to drag such a weight, I was totally at a loss to conceive. At this the Scot fired up, and wished to know "if I *dooted* his word." "Certainly not," I replied, "but still I am at a loss to know how he could hold her tail over his shoulder without its slipping through his paws." "Hoot, mon," said he, "dinna ye ken, that the canny old rat had tied her tail in a knot, and so had a fair hold of her?" This was a clincher; the whole company burst out into one roar of laughter, which so disconcerted our friend, that he seized his hat, and rushed out of the room, without bidding any one good night.

The "Quarterly Review" says, that rats will carry eggs from the bottom to the top of the house by lifting them from stair to stair. The male rat stands upon his head, and lifts up the egg with his hind legs, when the female takes it in her fore-paws, and secures it till he ascends a step higher; and so they pass from stair to stair, till they reach the top.

It is said that the ferocity and voracity of the rat surpass anything that can be imagined. A gentleman placed a dozen rats in a box, in order to try some experiments. When he reached home, lo and behold, there were only three, these three having gobbled up the other nine.

Now this tale is so imperfectly told, as to the time the rats were in the box, and the distance they were carried that we can make neither head nor tail of it; but, anyhow, it appears very much like hocus pocus, or high fly Jack quick and begone.

A few years ago, a gentleman living at Louth Hall, and taking a walk one morning round the establishment, in company with the coachman, the latter all at once drew his attention to the hen-roost, where a rat was struggling with an egg; and so intent was the animal on its undertaking, that it was perfectly unconscious of their presence. They watched its manœuvres for about ten minutes, and its principal difficulty seemed to be in balancing the egg; this it did, by stretching out one of its fore legs underneath the egg, and steadying it above with its cheek. When thus secured, it hopped very steadily and cautiously upon three legs; thus looking in its action more like a young rabbit than a rat. When he was near his hole, they gave him chace, and it was not till they were close upon him that he dropped his burden, to secure himself by running. The gentleman picked up the egg, and had it for his breakfast, and a very fine one he says it was. This, I think, clears up the mystery as to how rats carry eggs.

The following anecdote is upon the testimony of a very respectable and intelligent person, who insists most emphatically on the truth of her statement, because she herself was an eye-witness of the fact.

It appears that our informant, prior to her marriage, was a domestic in the establishment of a pastrycook, in London. Upon one occasion, her master had some very fine eggs, which he held in great estimation, and laid carefully away in the store-room above stairs. Presently some of these eggs were missing. Who had taken them? This was a question more easily asked than answered; for, until then it appears, none of the domestics knew anything about them. But how could they have gone? This was a greater mystery still, since it was quite clear that they could not run away of themselves. Nevertheless, with every search and inquiry, nothing satisfactory could be ascertained as to the disappearance of the eggs. At the same time it was reasonable to suppose, that though the scrutiny could not bring back those that were lost, yet it might tend to prevent others following them. In spite of all threats and precautions, however, the following morning proved to the contrary; for on inspection it was found that others had departed also. Now this was thought too bad: indeed, it was unpardonable; for if any of the domestics wanted eggs for their use, there were plenty of others, equally good, at their service, without taking those that were treasured like gold. This very naturally, and as might be expected, caused a great deal of uneasiness in the establishment, since every one felt equally innocent of such an act of wilful destruction, and at the same time were equally scrupulous in their suspicion of others. But on the following morning early, as the narrator of this anecdote was rising after a sleepless night, she thought she heard a noise on the stairs, and her first impression was of course that some one was stealing the eggs. She stole gently out of her room on to the landing, for the purpose of detecting the thief or thieves, if possible, and after pausing for a few moments, with breathless anxiety, to listen, she again heard the noise, but more distinctly than before. She cautiously peeped over the banisters, when she saw two rats, one large and the other small, most ingeniously conveying one of these eggs down stairs. It is needless to say, that the instant she saw them the whole mystery was explained, and her mind relieved of a world of cares. Indeed, so delighted and interested was she with the ingenuity and industry of the little vagabonds, that she offered them no molestation, but stood quietly gazing on with admiration till they were fairly out of sight. She says that when she first saw them they were both on one step, about half-way down stairs, when the big rat descended to the step beneath; he then stood upon his hind-legs, with his arms and head resting on the step above, till the other rolled the egg towards him; then putting his arms tightly round it, he lifted it carefully down on to the step where he was standing, and there held it till she came down and took charge of it, while he descended a step lower. Thus they passed from step to step till they were fairly at the bottom and out of sight. When the family arose, she explained the whole affair, to the great merriment and delight of the entire household; but more especially to her master, who, after laughing most heartily at the droll and ingenious contrivance of the robbers, concluded by assuring them, that he was perfectly satisfied with the honour and integrity of all about him, except the rats.

CHAPTER XII.

COURAGE, FEROCITY, AND CUNNING OF RATS.

In point of courage, rats vary considerably; for I have sometimes seen large rats exhibit such cowardice that they have yielded to their enemy almost without a struggle for their lives; while, on the other hand, I have seen both full-grown and half-grown rats fight most determinedly with anything that assailed them. But it must be borne in mind that there is a wonderful difference between poor rats that have been maimed in catching, and after that starving and pining in stinking cages, and those we meet with in their native haunts, brimful of health, strength, and vigour. Hence it is that many little bow-wows in London are lauded to the skies as invincible heroes for having killed some twenty or thirty half-grown, half-starved, half-dead rats. Whereas, take them into the country—to say nothing of sewer-rats—and let them take the rats as they rush from their holes, and I think that in most cases half a dozen would be more than these little pigmy heroes would care for, unless they were of a very genuine breed.

I once saw this matter put to the test. The owner of the dog was extremely loud in its praises, and certainly a very pretty little fancy animal it was, about six pounds in weight; but, nevertheless, its master had most wofully overrated its abilities, because it had once killed thirty rats in a pit, for which he bought it at a very handsome price. In the present instance he had backed it for a round sum to kill thirty fresh rats as they rushed from a bean-stack. Every one was requested to stand at a distance, so that the dog might have fair play. All being in readiness, and the servant waiting to bag the game, two ferrets were turned into the stack, when out ran a young rat; this he killed cleverly. The master called out to the servant to be quick and bag it; when out popped a second, rather larger, which gave him some little trouble, but he succeeded in killing it also. The smiling servant picked it up instantly, and popped it into the bag with the other. Then, after a few minutes' pause, out stole a big fellow. At him went the dog; but instead of his pinning the rat, the rat pinned him by the nose, and there held him till all his propensities for rat-killing had entirely evaporated in yells and squeals. Indeed they were forced to kill the rat to extricate the dog; and had they not held the dog tight, he would have tried who was the best runner. However, not all the threatening and coaxing resorted to could induce the dog to look at another rat, except to see that it was not coming after him; and so the master lost his money.

Ferrets, like rats, vary very much in courage. Five or six times have I seen ferrets so completely cowed by a rat, that they would never after look at another; and as for hunting, they would go anywhere you pleased, except into a rat-hole. But generally speaking, where there is no retreat, ferrets prove too much for ordinary rats; though my large tame ferret, of which I have already spoken, was so completely beaten by a single rat, that he would never after hunt anything. It happened, in my absence, that a neighbour brought a rat in a trap of his own construction, the ends of which came down with springs. It was a complicated affair; however, the rat was a very large one, though nothing near the size of the ferret; but one of the doors, or falls, had caught the rat by one of its hind-feet, and there held it tight. My father's porter brought the old ferret, and not only put it in at the other end, but shut it in. Upon returning soon after, and ascertaining what had been done, I opened the trap, when out rushed the ferret, bleeding and completely beaten; so much so, that never after would he face or hunt anything. The reason was obvious, for the rat having been caught by the hind-foot, had become so desperate, and being able to reach from side to side at the same time, from his position in the trap, he was compelled to stand fronting his opponent. The place, too, was perfectly dark; the consequence was, that whichever way the ferret turned, he felt the full length of the rat's teeth, and hence his total defeat; but could he have got behind, or sideways, the result might have been different.

Thus it is with inexperienced persons; they will spoil the best cat, dog, or ferret in the world! They forget that though these animals may have great courage, they have at the same time feelings the same as themselves, and are no fonder of being knocked about, or bitten by rats, than they are. However, my ferret was rendered worthless; and to warn my readers against another folly, I may here mention that this man spoiled a very excellent puppy of mine, by giving him a rat in a gin-trap; he did not let the rat loose, but threw it down, trap and all. The result was, that when the puppy shook the rat, he nearly knocked one of his own eyes out with a corner of the trap, and was blind ever after; nor would he ever touch another rat, or anything else.

A few years since, an account was published in the Irish papers, of a little girl who was passing through a field, when she observed a magpie and a rat engaged in mortal combat. In breathless haste she ran home, and brought some men to the spot, who found that, in her absence, a second magpie had come to the assistance of the first. The battle was raging furiously, with no flinching on either side. However, after some minutes of desperate struggling, they succeeded in killing the rat; though it appears he stood to his work most bravely, and fought to the last.

If this does not say much for his cunning and strength, it says a great deal for his courage; or he would have beat a retreat, and not stood to be picked to death.

It is a curious fact, that when male rats become aged and sombre, they will retire into solitude, and there live by themselves, in perfect enmity and disgust with the whole of their fraternity. In this way it is possible they may live a long time, and in the end die a natural death; whereas, by stopping among their community, they would most assuredly be eaten up as soon as they became drowsy, peevish, or decrepit. In their retirement they often fill out to a great size, through good living, steady habits, and plenty of rest; and will prove a sturdy opponent for any animal that may venture to assail them.

I was witness to a case of this kind, wherein my dog Pincher was most unceremoniously vanquished. We came inadvertently upon one of these grizzly gentlemen's retreats. He gave us no trouble to unearth him, but came forth with all the confidence and daring of a Celtic chief, to take a dignified survey of all around, as if to see who or what it was that dared to intrude upon his territory. He caught sight of Pincher, and with head and tail erect, pressed boldly forward to the fray. I cheered on the dog. The attack was mutual, but soon ended in the total defeat of poor Pincher. Still, I must remind you that Pincher, though a very handsome dog, was nevertheless a chance-bred one. Away he ran with all his might, yelping as he went, from the smarting effects of a deep wound he received just over the eye; and the conqueror retired doggedly to his hole, after making two or three threatening pauses, in apparent defiance of me. I could have killed him with my stick, but I was too much pleased with him. Nor did I feel justified, since he had won the battle fairly and honourably. However, I paid him a visit the next morning, accompanied by my little dog, Twinger! Twinger was a small, handsome, high-bred animal, not half the weight of Pincher, and possessing all the restless petulance and buoyant elasticity of a first-rate blood-horse. We were soon on the ground. Out came the rat to receive us, and seemed even more resolute than the day before. Perhaps his success with Pincher had emboldened him. Be that as it may, Twinger no sooner saw him, than at it they went, and a most desperate battle ensued. The ground was very sloping, being on a hill-side, and they rolled over and over several times; but the rat being so large, the dog's mouth was not big enough to grasp him; consequently the dog got most severely punished, which only made him the more determined. At last he succeeded in catching him across the nape of the neck, and then it was all over with the poor rat. Still he struggled boldly to the very last, and died like a hero.

Here was a degree of courage exhibited by a rat that I never saw surpassed by any other animal; for no animal, whatever its size or courage may be, can do more than fight till it dies, and this he did, with the most unflinching determination.

I may as well here mention that these old gentlemen, or rather retired rats, having been, as it were, surfeited with satiety, become entirely changed in their disposition ; and then possess feelings towards the females as loathsome and malignant, as before they were ardent and importunate. They will beat off or kill any rat, whether male or female, that shall dare to intrude upon their privacy. The consequence is, that they often, like the rats of Siam, keep the premises entirely free, where they chance to take up their abode ; and should any anxious mother seek an asylum in his domains, for the purpose of rearing her young in secret, as sure as he finds them out, so sure will he eat them, and her too, if he can catch her. Herein is he infinitely superior to either cat or dog, for keeping the foundation of your premises clear of vermin, because he can go where they cannot. But then it is necessary to be certain that he is a retired gentleman, and not a pioneer, before you allow him fair possession of your premises, or you may have cause to rue your indiscretion. Still this matter can be easily ascertained by a little careful observation. If he be regular in his haunts and feeding, and there are no signs of another about the place, then the matter is pretty clear ; but if you find that he allows others to keep him company, that instant set to work and get rid of them and him too, or you may soon be inundated.

About five-and-thirty years since, a countryman came to London, and brought with him a large tame rat, which he led about by a chain and collar ; but when the creature was tired he would carry it on his arm, or in his bosom or pocket. The curiosity of such a sight drew many pence from the passers-by ; nor indeed was this the only way the rat realized money for its master. It had fought many battles with dogs for various sums, ranging between five shillings and a guinea each match ; and in all of which he had proved victorious. But, alas! he fought one battle too many. It happened, that while standing by Tyburn Gate, with a number of people round him, a groom came along with a little white rough terrier ; but seeing the rat, he caught up the dog, and remonstrated with the man for letting the rat be on the ground. "Oh," said the countryman, "if you're not afraid of your dog, I'm not afraid o' my rat." The result was, that after some warm words, the animals were matched to fight for a wager of ten shillings ; and they retired to the grass in Hyde Park to decide it. The countryman no sooner took off the collar, and put down the rat, than the creature seemed quite ready for the contest. It appeared that Master Rat had one peculiar way of serving his opponents, which was to wait steadily till they made the attack, then seize them fast by the nose, and there hold them till they had had enough of it. The last battle, however, proved not only a failure, but the death of the little hero also. For when all was ready, the dog was let loose ; but instead of the rat seizing him by the nose, it missed its mark, and caught him by the side of the face. At the same instant the dog seized the rat by the lower-part of the belly, and tried to shake it, but he was too heavy. The countryman seeing the dangerous position of his pet, begged and prayed of the groom to take off the dog ; but before he could succeed the rat let go its hold, and fell back dead. The poor man burst into tears at the loss of his companion, and so pitiable was his lamentation, that it excited the commiseration of all around. The groom let him have the stakes for the dead rat, for the purpose of having it stuffed, and preserved as a trophy. Gentlemen and others gave him money, and even boys contributed their halfpence to sooth his sorrows. He talked to the dead animal in a most feeling manner. "Ah," said he, "we have travelled many a weary mile together ; and whether I laid in straw, or 'neath a hedge, thou hast always slept warm and comfortable in my bosom. But its all over with me now ; so I must e'en go back to the plough again, and settle quietly down to farm-work, or anything else they will give me to do, to earn an honest living."

In answer to various questions, he said he took the rat when very young, and carried it about with him wherever he went, and shared his victuals with it ; and from its becoming so tame and attached, he taught it many tricks, such as putting on and taking off a little cocked-hat ; also, at the word of command, to shoulder a little wooden gun, and go through various military manœuvres, such as ground-arms, shoulder-arms, right-about-face, quick march, &c. Then it would jump through a little hoop, or dance while its master whistled; and, lastly, it would finish its performance by running round after its tail, or rolling over on its back from side to side, to the great amusement of the lookers-on. After that it would stand on its hind-legs to beg; and the instant a coin was thrown down, it would pick it up in its mouth, and carry it to its master.

Some gentlemen, who had been listening to his narration, suggested that a rat was but a rat, and as he had been so successful in training one, why could he not train another? it would be but a work of time. This idea, it would appear, had never before struck him, and seemed to throw a bright ray of sunshine upon his prospects. The poor man smiled, and after thanking them most gratefully for their kindness towards him, went his way, evidently determined to try his abilities in the art of training another pet rat.

The "Quarterly Review" relates a circumstance wherein a sharp battle was fought between a game hen and a rat, in which the hen proved victorious by killing her grizzly opponent. But a gentleman living at Kensington, had previously experienced a very different result from a parallel case. A large rat paid a visit to his hen-house, wherein he had a number of fine Spanish fowls, and among them a favourite game-hen, which was the champion of the roost. It would appear, from the loud cackling and unusual disturbance among the fowls, and the quantity of feathers strewed about the place, that she had had a long and hardy struggle for victory, but was unsuccessful, since, instead of her killing the rat, the rat killed her. Great was the gentleman's disappointment on going to the roost; for instead of finding a number of new-laid eggs, he found the poor game-hen with her neck and shoulders dragged down a large rat-hole. On attempting to take her up, he felt that something was detaining her; so he gave her a good tug, when the rat, in his anxiety to retain the prize, allowed himself to be dragged half-out of the hole before he would relinquish his hold.

A gentleman told me of a battle between a rat and a Bantam cock, of which he was an eye-witness. He stated, that at his seat in Yorkshire there was a beautiful lawn behind, into which the dining-room opened by folding windows.

A friend of his, thinking that some Bantams would be a great addition to the place, and being himself a fancier of those tiny birds, made him a present of half a dozen, comprising five hens and a cock. They were of the most beautiful gold-spangle kind, known, I believe, as Sir Charles Sebright's breed. The cock weighed only a pound, and some of the hens less than that; and a very pretty sight he said it was to see them sparkling in the sun, or strutting about among the shrubs, or on the grass-plot. By-and-by one of the hens hatched nine diminutive chickens, and made as much fuss over them as if they had been six feet high each. However, the chickens were healthy and strong, and the cock was very proud of them.

When the chicks were about six weeks old, the gentleman missed two of them; on the following morning he missed two more; and the morning after that he missed another. He therefore charged his gun, and determined on watching the remainder of the day, and, if possible, shoot the paltry thief that could run away with such helplesss little creatures. The day passed on till after dinner without any sign of an enemy; but as he and his good lady were at dessert, they saw a large rat cowering among the shrubs, and evidently after the chickens; for every now and then he would creep nearer and nearer. The window was open when the gentleman cocked his gun and took his station, in order to let fly at him as he made a rush. Presently out he dashed, and, just as the gentleman was going to pull, the cock bird flew across his sight; and had he then fired he would have shot the cock and rat together; but, as it was, he laid down his gun to witness sport of another character; the cock was laying into the rat as hard as he could. He said the battle then became truly novel and interesting, though by no means a fair one; for if the rat once succeeded in getting hold of the cock, all further chance for the little fellow would be over. So he rang for the servant to bring his dog Wallace, a rough Scotch terrier. In the interim the rat got hold of the cock by the feathers at the back of the head. The cock fought on with all his might, when out came the feathers in the rat's mouth, which rather bothered him, because they stuck in his teeth. It now became the cock's turn; for the rat, in making a rush, missed his aim, when the cock seized him by the nape of the neck, and spurred him five or six times before he broke loose. The cock, now being winded, began to work round for breath; while, on the other hand, it was quite clear he had blinded the rat; for he reared on his hind-legs, with his paws up and his mouth wide open; and, after standing in this position for a second or two, he rushed furiously forward while the cock was behind him. These rushes he repeated three or four times, each of which the cock as dexterously avoided. At last the cock again attacked him. The rat turned upon him, and succeeded in catching him by the feathers of the thigh, just as the dog entered the dining-room. It now was quite evident the cock had no chance; so the gentleman let loose the dog, which instantly ran and laid the rat dead. For the dog's generosity, the cock jumped up and flew at him like a little fury, and followed him into the dining-room, when the gentleman drove him out, and shut the casement. His eyes flashed defiance, and he strutted up and down, and crowed enough to crack his little throat. He then ran off to the rat, and after pecking and striking at it several times, as if to wake it up, he crowed two or three times in token of victory. On the gentleman's examination of the rat, he found that both its eyes had been struck out. This affair certainly showed a great degree of courage in the little cock; but for the rat to continue the contest, even after it had been so painfully blinded, showed a degree of courage not often to be met with. Still this courage is not of that noble character which exerts itself for the mere love of victory, and will rather die than surrender; but it is prompted by the mere cravings of a ravenous and carnivorous appetite, or in self-defence.

The following is a curious account of a combat between a rat and a weasel, which appeared in some of the Scotch journals:—

"A man having entered a plantation on the estate of the Earl of Home, in pursuit of rats, he observed one in deadly strife with a weasel. How long they had been fighting before he came he could not say; but, being besmeared all over with blood and mud, they presented a sorry spectacle, parts of their bodies being rendered bare from the fur having been torn off. The man stood quietly to see the result, and he says that whenever they felt exhausted, they left off, and allowed themselves a few moments to recover their breath, and then set-to again with redoubled fury. In the last bout they continued closely engaged for about ten minutes with little advantage to either, though the rat was much the smallest. After that, the rat began to show symptoms of distress, and at last surrendered through sheer exhaustion, when the weasel twisted his body completely round that of his victim, and, seizing him by the throat, never left his hold until life was extinct. And then, as the man states, though much punished, he marched off with all the importance which victory alone inspires."

A few years since, as a washerwoman was engaged drying clothes on the dock at Dumfries, she accidentally struck her heel against an object on the ground, which proved to be a large rat. The infuriated animal turned upon the woman, and sprung at her throat with the most determined ferocity. She defended herself with her hands, and succeeded in knocking him down each time he attempted to fasten upon her. At last, in stepping back, she lost her balance and fell, when the rat attempted to follow up the advantage, and certainly would have done her some serious bodily harm, had she not instantly sprang to her feet, and renewed the contest. Again and again the little brute made for the throat, and the struggle might have continued long enough, had not another woman and a boy come to her assistance, when, by a liberal supply of thumps and kicks, they managed to lay him prostrate.

Another occurrence, unfortunately attended with circumstances of a melancholy character, took place on the premises of a livery-stable keeper, by which a gentleman's coachman lost his life. The evidence upon the inquest went to show that the deceased, in taking his horses into the stable, met with an opposition from the animals, which was very unusual, as they possessed no vicious propensities whatever. The deceased, in endeavouring to bring one of the horses up to the trough to eat its food, discovered a large rat in the manger, which, it appears, flew at the animals every time they attempted to approach, and hence their fright and reluctance to eat. The deceased, on making the discovery, endeavoured to kill the rat, which sprang over his head on the horse's back, and this so alarmed the animal that he plunged most violently; then rearing on his hind legs, struck down the deceased with such force as to break his jaw to pieces. The injuries were so severe that he died shortly afterwards.

I shall conclude this chapter on the courage, ferocity, and cunning of rats, by presenting two anecdotes, derived from authentic sources; the first from Mr. Jesse, and the second from the "Quarterly Review." Mr. Jesse, in the following narrative, gives a striking instance not only of the courage but the sagacity of these animals, as manifested in an experiment tried by a medical friend residing at Kingston, who is much devoted to the pursuit of natural history. "This gentleman," says Mr. Jesse, "having entertained a great deal of surprise and doubt that the ferret, an animal of such slow locomotive powers, should be so destructive to the rat tribe, determined to put the matter to the test, by bringing two of these animals fairly together, so that he might judge of their respective powers. He first selected a fine specimen of a large full-grown rat; then an equally strong male ferret—one that had been accustomed to rat-hunting. His son was with him, and they turned the two animals into an empty room which had but one window. Here they purposed waiting patiently to see the whole process of the encounter. Immediately upon being liberated, the rat ran round the room as if searching for a hole; but not finding any means of escape, he uttered a piercing shriek; then with the most prompt decision took his station directly under the light, thus gaining over his adversary the advantage of the sun. The ferret now erected his head, snifted about, and seemed fearlessly to push his way towards the spot where the scent of his game was strongest, at the same time facing the light in full front, and preparing himself with avidity to seize upon his prey. No sooner, however, had he approached within two feet of his watchful foe, than the rat, again uttering a loud cry, rushed at him, and in a violent attack inflicted a severe wound on the head of the ferret, which soon discovered itself by the blood that flowed from it. The ferret seemed astonished at the charge, and retreated with evident discomfiture, while the rat, instead of following up the advantage he had gained, instantly withdrew to his station under the window. The ferret soon recovered the shock he had sustained, and, erecting his head, he again took the field. The second rencontre was an exact repetition of the first, with this exception, that on the rush of the rat to the conflict, the ferret appeared more collected, and evidently showed a desire to get a firm hold of his enemy. The strength of the rat, however, was prodigiously great, and he again succeeded in not only avoiding the deadly embrace of the ferret, but also inflicted another severe wound on the head; he then returned to his retreat under the window, when the ferret seemed less anxious to renew the conflict. These attacks were renewed at intervals for nearly two hours, always ending in the failure of the ferret, which was evidently fighting to a disadvantage from the light falling full in its eyes whenever it approached the rat, which wisely kept its ground, and never for a moment lost sight of the advantage it had obtained in the onset. In order to prove whether the choice of his position depended upon accident, the medical gentleman managed to dislodge the rat, and then stationed himself under the window; but the moment the ferret attempted to make his approach, the rat, evidently aware of the advantage it had lost, endeavoured to creep between the gentleman's legs, thus losing sight of his natural fear of man, under the danger which awaited him from his more deadly enemy.

"The ferret by this time had learned a profitable lesson, and prepared to approach the rat in a more wily manner by creeping insidiously along the skirting, and thus avoiding the glare of light that had heretofore baffled his attempts. The rat still pursued, with the greatest energy, his original mode of attack, namely, inflicting a wound, and at the same time avoiding a close combat; whilst it was equally certain that his foe was intent upon laying hold of and grasping his intended victim in his murderous embrace.

"The character of the fight, which had lasted over three hours, was now evidently changed, and the rat seemed perfectly conscious of the wrong the gentleman had done it, in depriving it of its original advantage; at the same time, like the Swedish hero, it had taught its frequently beaten foe to beat itself in turn. At last, in a lengthened struggle, the ferret succeeded in accomplishing its originally-intended grapple; when the rat, conscious of its certain fate, made no further effort, but sent forth a plaintive shriek, and then surrendered quietly to its ruthless foe."

Mr. Jesse says, the progress of this experiment brought to proof the instinctive character both of the rat and ferret, and their reasoning powers. Still it must be borne in mind, that this was by no means a fair contest, because it took not only the ferret, but the doctor too, to defeat the rat; and I perfectly agree with Mr. Jesse's belief, that in a state of nature, or in a fair field, the rat would prove triumphant; but that in close quarters, and particularly in the dark, the insidious ferret would in most cases prove the victor.

For the authenticity of the next anecdote I am indebted to the "Quarterly Review."

A gentleman was witness to a battle between a rat and a ferret, with this great difference—the rat had fair play, and was allowed to hold fair possession of the advantageous position it had struggled for, and which position enabled it to beat off and defeat the wily ferret, which was absolutely bitten into shreds over the head and muzzle. The repetition of the same system of attack and defence by a second animal shows that this particular species of cunning is a general faculty of the rat tribe. Still, the main superiority of the ferret is, not only its dogged determination to retain its hold, when once it has fastened on its prey, but to suck the life's blood of its victim; whereas the rat mostly fights by a succession of single bites, which wound but do not always destroy. Nevertheless they will sometimes hang on by their teeth with great tenacity, and rather suffer death than relinquish their hold; still they never roll on their sides, and cuddle up their victim with that deadly, bear-like hug, as does the ferret, and the whole of the weasel tribe.

CHAPTER XIII.

UNITED ATTACKS OF RATS.

THAT rats feel unity to be strength, and becoming emboldened by numbers make united attacks and combined onslaughts, are facts not generally known or understood, but which the following accounts may satisfactorily prove.

At one end of the Serpentine River, there is an arch passing beneath the pathway to a precipice of rugged stones on the opposite side, for carrying off the superfluous waters of the river, after heavy rains, &c. On either side of this precipice there are trees, bushes, long grass, &c., in profusion; these form an excellent retreat for a multitude of rats, which have taken shelter there, and which, I am informed by one of the park-keepers, commit great ravages among her Majesty's ducks. On a summer's afternoon, as I was passing by this spot, there were numbers of these rats, of very large size, regaling themselves on the remains of a dead cat, which had been thrown over the railings. Several persons were looking on, when one individual pitched a little stone at them; but, so far from their being frightened or running away, they ran after the stone as it rolled down the precipice to the water at the bottom. This afforded sport for some minutes by keeping them running after small stones, when up came a boy with a little, active, barking cur; the boy hallooed his dog, and put it through the railings. When it made its way towards them, they all cowered down like so many cats after a bird, evidently with a design of making a simultaneous attack; but the dog, being suspicious of their design, began barking and jumping, as you will frequently see them do in front of a cow. The rats, nothing daunted, crept nearer and nearer, when one sprang at him, and another, then a third, and so on as they came within distance, all of which he as dexterously avoided by jumping out of their way and barking. At last the attacks and returns to the charge became so quick and determined, that, like a sensible cur, he turned upon his heels and ran away, evidently to the chagrin of the rats, who began quarrelling with each other. At this juncture, up came a dog-dealer with a very handsome black tan terrier in a chain. To unloose him was but the work of a moment, and in he went. The rats were ready to receive him. One sprang and met him, which the dog caught in his mouth, and as quickly let fall dead. A second sprang, and received a chop from his teeth and fell. A third had its life shook out of it, and so on, till he demolished some seven or eight, when away ran all the rest at once. The thing was but the work of a minute. Yet it was evident to every one present, that rats act in unison, and have their principles of attack and defence.

I was lately informed by a gentleman that in the house where he resided, the landlady had a very fine tortoiseshell cat, which was much valued for its abilities in killing rats, with which at times the premises were very much infested. The kitchens, it appears, were kept as mere lumber-rooms, whither the cat was in the habit of repairing, for the purpose of gratifying her propensity for rat-killing. However, one morning, when one of the domestics went down stairs, he found the poor cat in a most helpless and lamentable condition. The household was summoned to witness the catastrophe, and a pitiable sight it was. There lay the poor cat, perfectly helpless, and in the agonies of death. They may well say a cat has nine lives, for the flesh of the back and other parts of its body, from the nape of the neck to the root of the tail, was eaten entirely away to the bare bone. The poor animal was unable to move a limb, and looked up most piteously in their faces. My informant requested them, in charity, to put it out of its misery, which was done accordingly.

Now not one rat, nor do I believe half a dozen, could have so maltreated the poor cat, because she would have fought till she broke loose, and then have bounded up somewhere out of their reach. It must therefore have been a resolution on the part of numbers; for a cat is a most vicious and determined fighter in cases of danger; and nothing but overwhelming numbers could ever have served poor puss as she was served.

Not long ago there appeared in a country paper, the

account of a most extraordinary contest between a body of rats and some terriers. "The persons employed in some extensive grain stores, were one evening alarmed by the furious contention of dogs, and a great noise proceeding from the lower end of the yard. On the men arming themselves, and hurrying to the spot with lights in their hands, an extraordinary scene presented itself. There were thousands of rats gathered together on and about a cart, at the bottom of which lay the shakings of some meal-bags, and a furious battle was raging between the rats, on the one part, and two fine terrier dogs, belonging to the master of the stores, on the other. Emboldened by numbers, the rats firmly maintained their ground, though numbers of their comrades lay dead, while, on the other hand, the dogs were, as nearly as possible, overpowered when the men came to their assistance; and, even after the arrival of the men and lights, the rats did not retreat, but stood in obstinate defiance, till they were most unceremoniously beaten away with sticks and staves. The dogs were severely bitten all over their heads and bodies, and indeed in every part of them,—so much so, that on the following day they could scarcely move."

"This is the first case" (says the paper referred to) "in which we have known these noxious vermin to stand in combined force against dogs, though there are well-attested cases of their attacking children, and even grown-up persons."

Here we have a tolerably clear illustration of the union and combined determination, on the part of rats, to assist and support each other, in cases either of attack or defence; and there is little doubt, had the men arrived about a quarter of an hour later, that the dogs would have shared the fate of the poor tortoiseshell cat. Neither do I believe, had the celebrated Billy been there, or any of the renowned rat-killing dogs of the present day, that they would have fared any better; because overwhelming numbers will defeat anything. That there was no want of courage on the part of these dogs, the circumstances of the case prove beyond a doubt; for though they were severely punished, and nearly overpowered, when the men came, still they were sticking to their work like Spartans, and nothing about them in the shape of running away. But let us bear in mind, that these rats were, we might say, one family, and acting in unison; and not only that,—they were in their native grounds, and full of health and vigour. But how different is it with those rats that are put into the pits for matches! They, for the most part, are brought from various districts, and mixed together under most uncomfortable circumstances; consequently they have nothing like union among them.

It is recorded that, a few years ago, four condemned criminals made their escape from Newgate prison, by descending from a water-closet into the sewer, having formed the daring project of proceeding along it to the River Thames; but when they got as far as Fleet Market, they were beset on every side by such legions of rats, that the unhappy men screamed with agony. The people, hearing their cries, tore up the iron-gratings, and hoisted them out; when they were only too glad to be conducted safely back to prison.

The men who enter the sewers for the purpose of cleansing or repairing them, always carry lights, and go in too great force to be attacked.

A short time back appeared an account of a most ferocious attack made by rats on the wife of a labourer. She was in bed with her husband, and child, two years of age; when, about midnight, she was suddenly awoke by a most excruciating pain, and the agonizing shrieks of her helpless child. She threw up her hand, and grasped a great rat that had just plunged its teeth into her temple. The brute let go its hold, and after cutting her finger in two places, made its escape. Out of bed she jumped, and struck a light; then, upon raising her infant child, she found the blood streaming from several wounds in its head, as also from one of its fingers, which was fearfully lacerated. The whole was the work but of a few seconds. It was quite clear that a number of these ferocious creatures had made a simultaneous attack upon them; and had she not shown the resolution she did, the consequences might have been most disastrous.

CHAPTER XIV.

THEIR NATURAL WEAPONS.

HAVING treated of the courage, ferocity, and cunning of rats, we shall now proceed to notice those formidable weapons with which nature has endowed them, viz., their teeth.

The propensity of these animals for gnawing is not merely the result of a reckless determination to overcome impediments, but a necessity of their existence. "The rat" (says "Bentley's Miscellany") "has four formidable weapons in the shape of long sharp teeth; two in front of the upper-jaw, and two corresponding in the lower. These are formed in the shape of wedges; and, by the following wonderful provision of nature, have always a fine, sharp, cutting edge. On examining them carefully, we find that the inner part is of a soft ivory-like composition, which may be easily worn away; whereas the outside is composed of glass-like enamel, which is excessively hard. The upper-teeth work exactly into the under, so that the centres of the opposed teeth meet exactly in the act of gnawing; the soft part is thus being perpetually worn away, while the hard part keeps a sharp chisel-like edge; at the same time, the teeth keep growing up from the bottom, so that, as they wear away, a fresh supply is ready. Thus the ceaseless working of the rat's incisor-teeth against some hard substance is absolutely necessary to keep them down; and if he did not gnaw for his subsistence, he would be compelled to gnaw, to prevent his jaw being gradually locked by the rapid development."

In consequence of the arrangement of nature just described, there is a belief current, that if one of the teeth be removed, either by force or accident, the corresponding tooth will continue to grow, and, as there is nothing to grind it away, will increase till it turns upon itself, and forms a perfect ring. They also state, that if it be an under-tooth it sometimes will even run into the skull above. "Bentley's Miscellany" states, that there is a preparation in the Museum of the Royal College of Surgeons which illustrates this fact. It is, they say, an incisor tooth of a rat, which, from the cause above mentioned, has increased its growth upwards to such a degree, that it has formed a complete circle and a segment of another; the diameter of it, is about large enough to admit a good-sized thumb.

To substantiate this point, the "Quarterly Review" says, "We once saw a newly-killed rat to which this misfortune had occurred. The tooth, which was an upper one, had in this case also formed a complete circle, and the point, in winding round, had passed through the lip of the animal."

However curious all this may appear, and pretty in theory, still, in taking a practical survey of the subject, it appears very much like a physical impossibility. We know well that there is no accounting for the freaks of nature, and should she provide a rat with but one incisor tooth instead of four, then might we easily imagine that, from its curved or circular formation, it would, in the event of its continuous growth, and the entire absence of all obstructions, become a perfect ring, or rather a kind of diminishing circle. But for a rat, either by force or accident, to lose an upper tooth, and then for the corresponding one in the lower jaw to grow out of the mouth, and wind its way round till it passes into the skull, or for an upper tooth to grow through the under-lip till it forms a complete circle, seems to us not only mysterious but paradoxical; because, from the close proximity of the teeth in either jaw, and their flexibility, we cannot conceive how it is possible for a rat to gnaw a hole through our flooring, or through a brick wall, without the whole of the incisor teeth coming more or less in contact with the substance gnawed.

Now, with all deference to these gentlemen, let us, for experiment's sake, break out an under incisor tooth of a rat, and then keep him on soft bread and milk for six months, to prevent the teeth being ground down by any hard substance. What then might we expect? Why, that the sound pair, coming in constant contact with each other in the course of eating, would, to a certain extent, impede their growth; while, at the same time, the odd one, having nothing to check its growth, would extend to, say a quarter of an inch, beyond its next-door neighbour. Let us suppose this to be the case, and now let us give him a plank or the corner of a brick to gnaw, and what then will be the fact? Why, that the projecting tooth and the under one will grip the brick at oblique angles, while the corresponding one will remain idle; and, if the rat incautiously puts on too much power, the result will be that he will force the teeth out sideways, if not break them. But, in the event of his going more cautiously to work, the odd teeth will continue to act together till the protruding tooth is ground down to a slightly oblique level with its neighbour, and in this position will they remain as long as the rat lives.

This is our view of the matter; but as to whether the preparations spoken of are really rats' teeth at all, we shall leave them in the hands of more profound philosophers to decide.

It has often been matter of wonder and surprise as to how it is that both rats and mice can come head first down perpendicular fencings, trees, walls, wire-work, &c. without falling; but more especially that they do not tumble in, head first, and get a sousing for their trouble, when they attempt to drink out of cans, pails, tubs, &c. when the water is several inches from the top. I have watched both rats, mice, and squirrels, drink out of vessels where the water has been deeper down than the whole length of their bodies. The truth is, they can let themselves down and drink at the entire length of their heads, neck, body, and legs, as far as their claws or middle joint of their hind toes, and then escape without wetting a hair.

The reason is obvious, for if you examine their hind-legs and feet, you will find that they will turn outwards with the greatest suppleness till what we may call the heel is in front; so that the animals can hang on by their hind toes or talons to the bark of trees, rough palings, or dilapidated walls, and then let themselves down by a gentle shuffling kind of motion, at the same time checking themselves with the talons of the inside toes, or thumbs of the fore-feet.

As to their drinking out of deep vessels, I have seen both rats and a squirrel hang on to the edge of a pail with their hind toes, and quench their thirst when the water at the bottom has been scarcely an inch deep. But any earthen vessel, where the brim is thick, round, and glazed, bothers them. Hence arises the necessity for farmers having their milk-dishes either of glass or thick earthenware, well glazed within and without; and then they will preserve both their milk and cream free from their ravages.

However offensive and disgusting may be the appearance of rats'-teeth marks in our food or furniture, still to workers in ivory they are invaluable signs of excellence. The "Quarterly Review" tells us, that the rats are the best judges, and that the ivory-turners, &c. always look out for their teeth-marks on any part of an elephant's tusk they may wish to purchase, because the animals invariably attack that tusk, or portion of a tusk, which abounds with animal oil, in preference to that which contains phosphate of lime. Herein do turners, &c. find an infallible guide, where elasticity and transparency are the necessary qualities for a material much used in the decorative arts.

CHAPTER XV.

ARTICLES MANUFACTURED FROM RAT-SKINS,—LADY'S GLOVES, &c.

HAVING so far treated of the nature and habits of the rat, let us now proced to ascertain the nature of his skin and carcass.

Now my readers may be anxious to know for what purpose the London merchant purchased the six hundred thousand rat-skins in Paris; and for their enlightenment I will tell them. It was to change these skins from the loathsome, detestable things they were, into the most valued of all manufactured articles, because they are coveted by the ladies, and courted by the gentlemen. This may surprise the reader, but I will soon solve the mystery. In the first place, we will take the most beautiful, and at the same time the most dangerous, thing in creation. You will ask me what that can be, and my reply is—The fair hand of a lady, which, from the beginning, has made more fools of philosophers and sinners, of saints and sages, than all the temptations in the world besides.

"Now, Sir Charles ———" (to assume a case), "you had the honour, yesterday morning, of accompanying your betrothed to the mercer's shop, to procure some little necessaries for the evening ball. Those matters being settled, it then became your part to play the gallant knight; upon which you stoutly demanded to see some ladies' kid gloves—some of the very best French kid; for nothing less would suit. Down the shopman brought them with a gusto, being very sanguine in their praise, because he considered them an article not to be equalled; at the same time declaring he could fold a pair of them up and place them in a walnut shell. Your lady-love was so delighted, when she beheld them, that two pairs were instantly selected—a primrose pair for the morning wear, and a white pair for the coming evening. That done, you threw down your three half-crowns like a hero, and thought yourself the happiest man alive. But what followed? Why, you no sooner arrived home, than off the lady ran in ecstasies to mamma and papa, and sisters, and servants, and, indeed, any and every one in the house, to show them the dear, delightful present you had made her. Did the matter end there? Certainly not; for you, in the ardour of love, must have the honour of helping to try them on. You did so, and certainly nothing could be more purely white or delicately soft, except the fair hands that wore them! At least, so you thought. Then, with what a dim gaze of frenzy did you look upon the objects of your adoration, and to soothe the agitation of your throbbing heart, you sank as if by magic on one knee, then fervently pressing her hands to your quivering lips you kissed—what?—her hands?—No. French kid gloves?—No. What then?—Why, *rat-skins?* Nay, do not start; I say rat-skins, the primitive owners of which used to live by sometimes devouring horse-flesh at Montfauçon, and at others feeding on the veriest filth and garbage in the sewers of Paris; and yet withal, nothing in the wide world can equal rat-skins for beautifying even the fair hands of our most beloved sovereign Queen Victoria.

To me it is a mystery why rat-skins should be at all objectionable for gloves, when, at the same time, the skins of stinking polecats are sewn together and made into long rollypolies for twisting round the fair necks of our ladies, to shield them from the rude attacks of wind and weather, and which rollypolies are commonly known as boas. But to make the armory still more complete, a large round thing is made, somewhat resembling a drum, open at each end, to hide their hands and arms in,—which drum the furriers call a muff; and being covered with the skins of polecats, is pronounced to be *fitch* of the finest quality. But it is clear that fitch is neither more nor less than polecats' skins, into which the ladies thrust their delicate hands and arms up to their elbows! and with boas and tippets of the same material muffle up their ears and eyes, to make them warm and comfortable; yet they never dream of stinking polecats.

Then, again, the crowns of kings are rendered soft and easy to their royal brows by being lined and turned up with stoats' skins; but which stoat-skins are better known to ladies and furriers as *ermine*. In warmer and more temperate regions, the stoat is of a reddish-brown on the back and sides; and the throat, belly, &c. are white, except the tail, the tip of which is invariably black. But in colder climates, all but the tip of the tail is yellowish-white, and is held in great estimation for its fur. The consequence is, that in the icy regions, vast numbers of the native tribes, as well as foreigners, are in the habit of traversing immense tracts of country, amid ice and snow, for weeks and months together, in quest of these little animals. They are captured by hunting and trapping. Their skins and tails are torn off, and dressed, in order to keep the fur sound; and, when properly prepared, they are sewn together with the black tips of the tails sticking out here and there. This is the invaluable ermine which works so many charms and wonders among those teasing little creatures, the fair sex! Still, at the same time, what more delightful or dignified present can any lord make his lady than presenting her with a complete suit of ermine, comprising muff, cuffs, cape, tippet, boa, and cloak. I know of no present that will give such general and infinite satisfaction; and if any one doubts what I say, let him try the experiment with his wife or betrothed, and see if they do not in return declare him the dearest creature living, and become as loving and affectionate as the comfort and warmth of stoat-skins can make them.

How comes it, then, that polecats' and stoats' skins are held so inestimable, while the poor humble rat's skin is held in detestation, when in texture and softness it is quite equal, if not superior, to either? The reason is obvious. Over the whole surface of our island, from shore to shore, east, west, north, and south, there is scarcely a hole or corner in our homes and farmsteads that is not infested with rats; while the banks of our lakes, rivers, streams, and ditches, are completely drilled and intersected with their subterranean runs and retreats. Here, then, is the objection—they are too common. Still I am satisfied there is no one thing can equal them for ladies' gloves, where delicacy and softness are the essential requisites to form the beau-ideal of perfection. But my father's maxim may perhaps better explain the mystery. He always maintained that anything far-fetched and dear-bought was good for the ladies. Nevertheless, in despite of every prejudice, should my readers at any time wish to impress those delicate things, the ladies' hands, with the pure stamp of elegance, just thrust them into rats' skins, and the work is complete.

In order that the gentlemen may not laugh at the ladies' expense, allow me to remind them, that if the ladies wear rats' skins upon their hands, the gentlemen beautify their heads with rats' fur, and walk along with as much importance as if they wore the coronets of kings. The "Quarterly Review" informs us that in France, the sewer authorities hold an annual hunting-match, on which occasion there is a great capture of rats. These animals are not destined to afford sport under the tender mercies of a dog "Billy, Jem, or Tiny." On the contrary, the French have too much respect for the soundness of their hides. Then again, a company has established itself in Paris, on the Hudson's Bay principle, to buy up all the rats of the country for the sake of their skins. The soft nap of the fur, when dressed, is of the most beautiful texture, far exceeding in delicacy that of the beaver; and the hatters consequently use it as a substitute.

But to carry out my point, Sir Charles ——, let us turn to the grand ball. Last evening saw you, in the etiquette of fashion, dancing with the rival of your intended. What unconscious cruelty were you then perpetrating upon one who doated on you—upon one who, amid the gorgeous splendour and glittering throng, saw nothing but you—and you dancing with her rival. Fie upon you! With what finished grace you led the hated one to the dance; then how you exerted every nerve to round with ease and elegance your every action, by which you thought you were making rapid strides in the affections of your chosen one; but which display was, in reality, cramping her very heart with jealousy. Talk of etiquette; I believe my good old Yorkshire grandmother's etiquette, if not the most fashionable, was at least more natural; she was eighty-two when last at a ball, and she then declared that my grandfather, who was three years older than herself, should never dance with any one but her while she was living; and, what is more, she kept her word, and was proof to the last against all appeals. It may appear quaint or even vulgar in me, but I cannot help thinking she was right, when we see so many miserable consequences arising out of this first familiarity in a ball-room.

Unconscious of all around, there, Sir Charles ——, sat the idol of your heart, biting the rat-skins till she had laid the ends of her fingers bare. At that moment, sir, she was a perfect little cannibal, and would have bitten you and her rival to pieces with more avidity and less remorse than she did her rat-skin gloves. Presently she quitted the room, and left for you a card, requesting you not, on any account, to call upon her till this evening. Now, sir, I would sooner be fastened in a barn with five hundred rats, were I you, than be locked in a room with this pretty little cannibal. Indeed, sir, I would rather present her with fifty pairs of prepared rat-skins, carcasses and all, than listen for half an hour to one of her reproving sermons. Nevertheless, were I in her place, as a punishment for your thoughtlessness, you should swallow the remainder of the gloves—both pairs I mean—before I would pardon your seeming perfidy.

There was one interesting circumstance, connected with the ball-room, which I have yet to mention. While your little white rose, Sir Charles ——, was quietly eyeing your every movement, Lady Rattle, in the most courteous and inquisitive manner advanced, and thus addressed her:—" My dear Emily, you are quite the belle and idolized deity of the room. You are turning them all heathens and idolators; for the whole adoration of both sexes is centred upon you. Now, dear, for Heaven's sake, do undeceive them, or you will be answerable for their eternal loss!" Emily inquired in what she was to undeceive them? "Why," Lady Rattle replied, "to tell you the truth, love, I am deputed by some of them to ascertain whether you have on gloves of a supernatural quality, or whether it is the exquisite delicacy of your skin which so dazzles and bewilders the whole company; for nothing is either seen or spoken of but your hands, dear." Emily sedately informed her that she had on gloves which were a present from Sir Charles, and that she believed they were a new kind of French kid. "Dear, dear," said Lady Rattle, "what a little delicate creature it must have been! No doubt it drank nothing but new-milk, and was fed upon cream and almonds. How delicious it must have eaten! Oh, how she should like to have lunched off it! Its flesh must have been as white as a curd. Do you know, love, I'm going to give a superlative treat, and I must insist upon your coming to it. I intend giving a real French kid dinner to all my friends and acquaintances; there'll be a sensation for you; 'twill be the leading topic of the day; and I shall employ M. Soyer to serve them up. Sir John is going to Paris next week; and to prevent imposition, dear (for there are no real French kids in England), he shall bring over a number with him; and, oh, we will have such a treat, love! For, do you know, dear, that French kid-soup, or stewed French kids, impart such a transparent delicacy to the complexion, that nothing on earth can equal it! Oh, heavens! we shall have all the gentlemen dying in love for us. There'll be a conquest! Well now, dear, I must pray you to let me go, for poor Sir John is languishing for me." They then parted.

CHAPTER XVI.

RATS AS HUMAN FOOD.

In treating on a subject so generally viewed with feelings of disgust, let us, by way of preliminary, only imagine Sir John bringing over from Paris a cage full of sewer rats, for the entertainment of Lady Rattle's friends and acquaintances; and that no less a personage than M. Soyer, the celebrated Crimean chef, should be employed to kill, skin, cook, and serve them up to table. But from imagination let us turn to the realities of life.

Strange and revolting as the above might appear, yet if Lady Rattle had had rats for dinner, she would not have been singular; for, if chroniclers and travellers report truly, there are very large portions of the human family who esteem rats not only as a necessary of life, but a great delicacy. Among them there are the whole of the Chinese, and many natives of the East Indies, besides the tribes of Africa, and the natives and more civilized freed slaves at Sierra Leone. Then there are some islands in the Pacific, where the only animals the natives have to live upon are pigs and rats. Then we find that the humbler classes and lazaroni of Naples, and the chiffonniers of Paris, indulge in them as an article of food; and, to bring the matter nearer home, there is Mr. Wilkie Collins's account of the good people of East and West Looe, Cornwall; besides some persons at Hanwell, and a host of individual cases too numerous to mention.

My readers may wonder what I am driving at, and whether I am going to set them to eating rats? In reply, they will allow me to tell my tale my own way, and we shall arrive much sooner at a conclusion. There are some things in this world rendered so hideous by misconception, that to expose them suddenly to the caprices of imagination would be to bring on an instantaneous fit of swooning; but as I have no desire to see such unhappy consequences, I shall strive to unfold the various subjects with all the modesty and gentleness that a blushing rose opens amid the spangling dews on a morning in June. Yet, as the completion of my undertaking compels me to a mature consideration of the present point in Ratology, I must crave a little leniency or indulgence, as I am about to handle perhaps the most loathsome subject in the whole range of human dietetics; and I wish to do it so as not to disturb even the most delicate stomach or squeamish antipathy of the most sensitive or fastidious of the human family.

There was a curious anecdote current some few years since, to the effect that, at the conclusion of the Chinese war, the British admiral, prior to his departure for England, was invited on shore to dine with Commissioner Lyn, one of the Chinese dignitaries. No sooner had he and his officers arrived, than the dinner was served up; and the admiral and the commissioner were seated on opposite sides of the table, according to European custom; but the interpreter, for some reason not assigned, was absent. The covers were removed, and all was steaming temptation. The commissioner smiled, and rubbed his hands with evident satisfaction; but not so the admiral, for his mind seemed brimful of reflective considerations. Nevertheless, being very hungry, he helped himself most unceremoniously to half the contents of a dish in front of him; and, after eating the meat, and picking the bones with the utmost relish, he quaffed a goblet of wine, then plunged his fork into the remaining half; when, all of a sudden, his attention was arrested as to what he was eating. This brought him to a dead stand-still; while sad misgivings rushed through his mind as to whether it was a dog, duck, or dolphin. Still he never recollected having met with leg bones in a fish before—then it was quite clear it was not a fish. What then was it? a beast or a bird? There was the problem! and from his sunken expression of countenance, it required instant solving; not so much to give his stomach a relish for the remainder as to retain what he had already eaten. But how was the matter to be explained, since the commissioner and the admiral could not understand each other in a single word, and unhappily the interpreter had not yet arrived. What was to be done? Here the admiral's ingenuity, to a certain extent, overcame the difficulty; for to make himself understood he had recourse to the following admirable stratagem. "Sir," said he, at the same time holding up the thing in question upon his fork, "is this the remnant of a 'quack, quack, quack?'" to which question the commissioner significantly shook his head, as much as to say that it was not a "quack, quack, quack," but said it was a "bow, wow, wow." The astounded admiral was not willing to believe his own ears. Indeed he strove, by every means in his power, to persuade himself that the commissioner was either joking, or did not understand him; therefore, to satisfy himself, he asked more slowly and emphatically than before, if it were a "quack—quack—quack." To which the smiling commissioner again shook his head, and replied with equal distinctness, "bow—wow—wow," which made the admiral look grave and serious, as if he thought it was a matter rather derogatory to the honour of England for a Chinaman to poke fun at a British admiral. He suddenly grew warm, and in a guttural tone said, "Confound the fellow! is it a 'quack, quack, quack?'" When the commissioner, with equal warmth and quickness, snappishly replied, "bow, wow, wow." They then, as with one impulse, rose to their legs, as did also the attendants on both sides. The admiral stamped and roared, while the commissioner jumped and screeched. At last they became so enraged with each other, that they roared, stamped, jumped, and screamed, till nothing was heard but a hoarse "quack, quack, quack," and a screeching "bow, wow, wow." At this moment in came the interpreter, who soon brought the floundering admiral to an anchorage; for no sooner had the admiral caught sight of him, than with clenched fist upon the table, and in a voice of thunder, he demanded an instant explanation as to what he had been eating. To which demand the interpreter complied, by telling him, in the most bland and conciliating manner possible, that it was a magnificent roast-puppy, the offspring of Lady Lyn's favourite lap-dog, and which the commissioner himself had that morning purloined to do honour to the occasion. The admiral stood aghast. It is needless to say he had quite finished his dinner; and for decency's sake rushed out of the room. This extraordinary departure so astonished the worthy commissioner, that he stood perfectly bewildered, till the interpreter fully explained the nature of the case; because, till then, he had thought this uproar or "filliloo" at table was only an English custom, and had felt but too delighted with the comical idea. At the same time he had exerted himself most anxiously to indulge the admiral in this supposed national propensity.

Presently, with swollen eyes and haggard aspect, the trembling admiral returned to bid his generous host farewell. The penetrating eye of the commissioner saw that his frame had been undergoing a severe agitation, and, in soothing accents, for his stomach's sake, he bid the interpreter to invite him to another entertainment, when he would take care there should be no nasty "bow, wow, wow," but plenty of nice *cat, cat, cat,* and some nice *rat, rat, rat!* The promise of the last was a perfect settler; for had they not instantly supported him, he would have fallen. He was, with all speed, conveyed to his ship, still suffering under the agonies of a heaving stomach; nor was it till after he had been seated for half an hour behind a salt buttock of beef that he felt anything like composure. Then, after taking sundry noggins of brandy, a genial glow of indescribable tranquillity stole through the "ports and alleys" of his system, and gently wrapped him in the soothing arms of Morpheus. The ship's doctor sat up all night to watch him, and towards morning he began to show symptoms of uneasiness. This uneasiness increased, till at last he raved out curses and imprecations without measure upon all the rats, cats, and dogs in the Celestial Empire. With this he awoke; then, after collecting his scattered senses, and ascertaining that the wind was favourable, he gave orders to weigh anchor, and instantly set sail for the land of roast beef and plum-pudding.

Now, whether this tale be founded on facts or whether it be purely fiction, it matters little; but yet it is singular to read with what force the prevailing antipathy to the bare name of *rat* is turned to account. For when the admiral was informed that he had eaten half of Lady Lyn's favourite puppy, it only turned him sick; and when he was promised some nice cat, cat, cat, he merely shuddered; but the instant the word *rat* was mentioned, his legs gave way under him; thus possibly depicting the unqualified disgust and horror that is entertained against the mere name of the animal about which I am writing.

Shakspeare has said, "What's in a name? for that which we call a rose, by any other name would smell as sweet." So far so good. But with all due deference to the immortal bard, I maintain that rats by any other name would *eat* much better.

Here, I think, in justice to this little animal, we ought for a few minutes to divest our minds of all unpleasant associations, and then see how far it has fair cause of complaint against man, in reference to the odium with which it has been branded. After this, show me, if you can, one animal, in the whole range of British zoology, that is cursed with so foul a name; or show me, in the vocabulary of the English language, one word that is so harsh, grating, and displeasing to the human ear, as is that of *rat*. Just spell it slowly, and then pronounce it, R—A—T, rat. There is such a rattling at the tip of the tongue; and then its sudden and abrupt termination with T, reminds us of a bolting horse coming smash against a turnpike-gate, and being thereby thrown on its back. There is nothing to modify it. Indeed, of all words in the English language with which I am acquainted, it is the most discordant and ungrateful in its sound, and conveys an impression of something indescribably foul and offensive to the imaginative eye and ear—something to be avoided at all hazards. Now, imagine for one moment, that you are a native of the Arctic regions, and that you read, or hear, for the first time, that the people in England are terribly infested with a thing called—a rat. Then your mind would conceive a something which was scaly, filthy, and mangy in the extreme; with a form hideous and undefinable, and emitting a loathsome, pestiferous odour that poisons all around—a voice unearthly, and a precursor or harbinger of death; in a word, a kind of devil's lapdog, that had been kicked out of the infernal regions for being too offensive and too ugly, but which has an everlasting craving for men's hearts or ulcerated toads.

This you may think is something rather overdrawn; but the best way to test it is to imagine such a monster newly thrown amongst us; or, if you can conceive something even more hideous still, do so, and then see if you can find a name more appropriate and indicative of its many malformations than that of rat. Whereas, in truth, the rat is no such thing; but, on the contrary, with all its depredations, the farmer's rat is one of the most pretty, cleanly, active little animals that ranges our dwellings, fields, and forests, and possesses a skin even more soft, glossy, and flexible than the best Genoa velvet. And as for the disposition of the young, if you saw them around a rick on a moonlight night, or a bright summer's morn, skipping and playing, and going through a variety of little gambols, with all the cheerfulness and innocence of lambs or kittens, you would scarce believe your eyes, or that they were in truth the so universally-despised creatures which, in England, every man, woman, and child holds in such fear and abhorrence.

Yes, you say; but look at an old sewer-rat! This is not a fair view; for take any ill-used, ill-housed creature in the winter of life, and what shall we find? Why, that the bloom of health is sapped and faded—the fire of the eye burned down to the lingering embers—the vigorous elasticity and buoyancy of youth grown into fretful and restless decrepitude, and a general declining of the whole being to that "bourn from which no traveller returns."

Can you then tell me what there is more condemnatory in a rat growing rusty with age, than in any other creature doing so? But let us proceed.

It appears to me a sound logical conclusion to arrive at, that if in France rats' skins are French kid, then their carcasses must of necessity be French kids; or if you prefer it we will say Norwegian kids, or Hanoverian kids; or perhaps it may suit you better to call them British kids. Then let us see how the despised rat-catcher will be affected thereby in the scale of estimation; for, instead of the despicable title he now bears, he would be styled the French or Norwegian kid-hunter; while others would maintain that he was a Hanoverian or British kid-hunter. Now would not either of these imaginary titles exalt him into universal admiration, and at once place him in bold comparison with that noble race of mountaineers, the chamois-hunters of Switzerland? I think it would. At the same time, would not the animal itself be divested of nearly all its terrors, and be held in the same light as hares, rabbits, squirrels, weasels, &c.? We need not, however, trouble ourselves to find it a more agreeable name than the odious one it has held so long, since the great French naturalists, Count Buffon and Baron Cuvier, have already supplied it. They call it the "*Sermulot.*"

Here then is a name which is as smooth and pleasing in its sound as the other is harsh, grating, and discordant. Therefore, let us for the remainder of these articles style it the "Sermulot."

The author of the popular "History of Mammalia," says there are parts of the world besides China where sermulots are eaten, and such sermulots, or rats, as would astonish those accustomed to the British species, which, take even the largest, are Liliputians, when compared with a native of the East Indies, as first satisfactorily described in the "Linnæan Transactions." The specimen here described was a female, and weighed two pounds eleven ounces; its total length being two feet two inches and a quarter. It goes on to state that the male grows larger, and weighs three pounds and upwards; so that the natives have on the table before them an animal as large as a wild rabbit; and, doubtless, as they have no prejudices or scruples, the animal eats just as palatably. It is also affirmed that the quality of the flesh is according to the food it eats.

In a work entitled "Three Years' Residence at Sierra Leone," the author states, that the more civilized among the liberated Africans esteem the rat (or rather, as they call it, the ground-pig) as very good food, and that the author's domestic servants constantly set snares and traps for the rat and other bush meat, such as squirrels, monkeys, &c.

Now, which of the two would be most pleasing to the eye, and consequently eat the best,—a ground-pig or a monkey? I here speak upon the testimony of others, that when properly bled and prepared for cooking, no diminutive sucking-pig, however choice, can look more fair, plump, and tempting than does the barn-sermulot, with its little kidneys enveloped in a delicate bed of pinky-white fat; while the monkey, on the other hand, looks like a little shrivelled-up Indian.

A lady informs us, that when at Naples she was surprised to see, in some of the shops, rows of sermulots suspended by their tails like ropes of onions for sale. Upon inquiry, she was informed that they were eaten by the lower order of

Neapolitans and the lazaroni. The journals also inform us that there are over 150,000 Chinese in California, who make excellent farm-servants, and work very cheaply, because of the size, quality, and numbers of the sermulots with which the country abounds.

The "Quarterly Review" tells us that split dried sermulots are sold in China as a dainty; that the chiffonniers of Paris feed on them without reluctance; and that sermulot-pie is not altogether unknown in our own country. The gipsies, the writer states, continue to eat such as are caught in stacks and barns; and a distinguished surgeon of the present time frequently has them served up at his table. He also mentions that the British navy was not always so nice as at present; for an old captain in Her Majesty's service told him that on one occasion, when returning from India, the vessel was infested with sermulots, which made great ravages among the biscuits; and therefore, to make up for the loss, the sailors used to catch all they could, and put them into pies, which they considered a very great delicacy; and lastly, he tells us, that at the siege of Malta, when the French were hard pressed, sermulots fetched a dollar a piece; but two dollars was given for those caught in barns, ricks, or granaries.

We have a full and clear account recorded by Mr. Wilkie Collins, in his "Foot Rambles in Cornwall," how the good people of Looe got rid of their ground-pigs or sermulots. He says that about a mile out at sea, there rises a green, triangular-shaped eminence called Looe Island. Here several years since a ship was wrecked. Not only was the crew saved, but several free passengers of the ground-pig or sermulot tribe (which had got on board, nobody knew how, where, or when,) were also preserved by their own strenuous exertions, and wisely took up permanent quarters on the land of Looe Island. In process of time, and in obedience to the laws of nature, these ground-pigs, or sermulots, it appears, began to increase and multiply so fast that they soon became an intolerable nuisance.

Their numbers not only threatened destruction to the agricultural produce of the island, but it became a matter of great doubt as to the safety of a man going by himself on the island, lest he should share the fate of Bishop Hatto, and be eaten alive. The ordinary means of destruction had been resorted to, but with trifling effect. How then was the impending ruin to be averted? Here was matter for most serious deliberation. The result was, that a meeting of the cleverest heads of the town was convened to sit in council, and give judgment according to the evidence adduced. This went to prove that not only had a horde of ground-pigs or sermulots taken entire possession of the island, but that after they were laid on the ground for dead, they, by some means unknown to any but themselves, recovered faster than they could be skinned; then making their escape into their holes, they seemed to increase and multiply faster than ever.

The great question now at issue was not how to kill them, but how to annihilate them so effectually that the whole population might know for certain that the reappearance even of one of them would be a thing beyond all possibility. Here was the problem; and which was practically and triumphantly solved in the following manner:—A grand hunt was proclaimed for a certain day, to be followed by a splendid entertainment for all the huntsmen as well as their wives, children, and friends. At the time appointed all the enterprising spirits of the town sallied forth in a body to Looe Island; and great was the destruction among the sermulots, which were as speedily conveyed across the water to the inn, and thence to the kitchen, where the cooks soon found themselves up to their eyes in business, preparing them for dinner. The throng of company met in good time, and after being seated in due order around the various tables, the sermulots were speedily served up, steaming in large clean china dishes, and smothered in onion-sauce. At the given signal, every cover was removed, and the sermulots, says Mr. Wilkie Collins, were straightway eaten with the keenest relish by the good people of Looe.

Never in England was any invention for destroying sermulots so eminently successful as this. Every man, woman, and child present could swear to the death and annihilation of all the sermulots they had devoured; and such was the liking thereby created for this species of game, that the local returns of dead sermulots were not made by the bills of mortality, but by the bills of fare. Day after day were they hunted for public and private consumption; and thus did they disappear by hundreds, never to return.

If we take a walk across the country to Hanwell, we shall there find a good-looking young man commonly known as "Shotty," and who is in partnership with another man living at Ealing. These individuals inform me that they are the principal sermulot-hunters throughout that part of the country; that is to say, within eight miles round Hanwell, comprising an area of about one hundred and forty-four square miles of fertile country, thickly populated with sermulots; yet they cannot live by their legitimate profession. Still, if you feel sportively inclined, just venture a shilling, and "Shotty" will prepare, fry, and eat a sermulot before your eyes, and thank you for the shilling when he has done. Indeed, the last time I saw him, he said he only wished he could get a shilling each for all he had eaten, that then he would leave off work, and set up as gentleman for the rest of his life.

Now I should like to learn what just charge there is to be preferred against the flesh of the barn-sermulot more than against the flesh of hares, rabbits, pheasants, and partridges. They eat the farmer's corn, and what more wholesome feeding can anything have? and as to their domestic habits of cleanliness, they stand second to none. And though my own stomach has its loathings and misgivings, still that argues nothing but prejudice and antipathy.

Nevertheless I am fully satisfied, in despite of every weakness, that to those who can like them, and eat them with a relish, they will prove not only wholesome, but a cheap and nutritious meal.

The great question at issue appears not as to the propriety or utility of introducing cooked sermulots, either at public or private entertainments, but as to how far their flesh may be wholesome or unwholesome as food for the human stomach. In addition to about one half of the human family, we find that birds of the eagle, owl, and falcon tribes, as well as all kinds of carnivorous animals, eagerly pounce upon the sermulot, when they can do it with safety, and eat it with the utmost relish and avidity. Pigs, pike, poultry, and the larger eels, will eagerly regale themselves with sermulots when an opportunity offers. Still I never in my life heard of either man, bird, or beast being poisoned, or even made sick through swallowing a sermulot! But I have met with numerous individuals who have eaten them; and one and all, without an exception, have declared them to be most delicate and sweet eating; and in many cases they have given them preference to rabbits.

CHAPTER XVII.

PREJUDICES AND ANTIPATHIES.

PREJUDICE and antipathy, forsooth—certainly the veriest limping, blinking, one-eyed, one-sided bigots that can benumb the heart, and poison the springs of human sensibility, but which shrivel into the shade before the light of truth and common sense. Nevertheless, there have been eminent individuals in all ages who have been the victims of some ungovernable antipathy, which was perhaps at first only a dislike; but being fostered and cherished through a period of time, had grown into an aversion, and ultimately became incorporated with their very nature, so as to resolve itself into a deep-rooted antipathy, which nothing but the incidents of cruel misfortune, or the iron hand of necessity, could ever shake or overcome.

Sir Nicholas Bacon, Keeper of the Great Seal and Privy Councillor to Queen Elizabeth, and who was the first Keeper that ranked as Lord Chancellor, with all his learning and dignity, would swoon away at each eclipse of the moon; and the insolent but clever Scaliger, of the sixteenth century, is said to have trembled in every limb at the sight of water-cresses.

The celebrated astronomer, Tycho Brahe, of the sixteenth century, nearly expired at the sight of a fox; and the Duke d'Epernon would swoon on beholding a leveret, although a hare did not produce the same effect. Henry III. of France would also faint at the sight of a cat; and Marshal d'Albert was affected in the same manner at the sight of a pig.

Vladislaus of Poland would run from the sight of apples; and Montaigne, remarking on this subject, said that there were men who dreaded apples more than cannon-balls. It is also stated that Peter the Great, who contemplated the subjugation of the world, would flee from any house on beholding a black beetle, and that no persuasion could ever after induce him to enter the place again.

The learned Ariosto would shudder at the sight of a bath; and Jerome Cardan, the great Italian physician and astrologer, would also shiver at the breaking of an egg.

Orfila relates an instance of a lady, forty-six years of age, who could never be present when linseed-tea was preparing, without being troubled, in the course of a few minutes, with a general swelling of the face, followed by fainting and a loss of her intellectual faculties, which condition used to last for four-and-twenty hours. Zimmerman also tells us of a lady who could not endure the feel of silk or satin, and shuddered when she touched the velvety skin of a peach.

That most learned scholar of the sixteenth century, Erasmus, who came to England by the invitation of Henry VIII., always took a fever when he smelt fish; and that daring warrior, Julius Cæsar, whose towering ambition grasped at nothing less than the whole world, would tremble in every limb at the crowing of a cock.

Voltaire gives the history of an officer who was thrown into convulsions, and lost his senses, by having pinks in his chamber.

Orfila gives the account of the painter Vincent, who was seized with violent giddiness and swooned when there were roses in the room.

Mary of Medicis and Cardinal Gardona, whenever either of them smelt the odour of a rose, would be directly seized with a fit of sickness; and it is said of the poet Lord Byron, that he felt the most perfect horror and disgust at seeing ladies eat; and I myself have seen a long, lean, hungry-looking fashionable at table, refuse a fine smoking sirloin of beef, because (to use his own words) he was not beef-hungry. I have also seen a similar-looking customer at a tavern refuse a fine rump-steak, because there was neither tomato nor oyster sauce with it. On another occasion, I heard a fine fellow bid the waiter give a splendid mutton-chop to the dog, because there were no pickled walnuts; and then I saw a highly-starched fellow upset the happiness of a merry gipsy-party by refusing to eat his dinner, because the ladies had brought a piece of good old Cheshire cheese to eat after it instead of ripe Stilton.

Now, do you not think that if each of these worthies had been supplied with some fine curried sermulots, they would have enjoyed them vastly, and, after using their silver toothpicks, called for their sherry to do honour to the fragrant and delicious repast? Indeed, to hear some of these tavern aristocracy talk, one would imagine them to be of such exquisite alabaster composition, that to feed them on anything meaner than liquidized crystal, iced creams, and syllabubs, would be to perpetrate a crime, which in itself would be so revolting, that "murder most foul," and Herod's massacre of the innocents, would sink into mere child's play and holiday merry-making; and as for poor old Epicurus, he, as a matter of necessity, sinks into a mere omnivorous old cannibal.

But necessity, which is said to be the mother of invention, is also the infallible curer of all foolish prejudices and antipathies. For I find, in Archer's "Vestiges of Old London," some extracts from Stow's "Annals," wherein he records, that in addition to the numerous visitations of fire and pestilence, the annals of London include repeated instances of famine, the most frightful of which is recorded as having occurred in the beginning of the fourteenth century, when London, in common with the rest of the kingdom, was visited by so complete a famine, that the wealthy esteemed horse-flesh a complete luxury, and the poor fed on rats and mice, and would relish a stale dead dog that died of starvation. Some, who were secreted in hiding-places, were, like the mothers of Jerusalem, compelled through hunger to eat the flesh of their own children, and would devour any stale carcass they could meet with. The thieves that were in prison would tear in pieces those that were newly brought amongst them, and greedily devour them half alive.

Now, into what perfect insignificance do all the various antipathies sink when contrasted with the horrors of famine. In a word, when horse-flesh was esteemed a complete luxury by the wealthy, pray what amount of money would not a fine hot sermulot-pie have realized?

CHAPTER XVIII.

WHISTLING JOE, THE HERTFORDSHIRE SERMULOT-HUNTER AND RAT-CATCHER.

SOME few years ago one of the best customers of a well-known tavern at Hertford, was a celebrated sermulot-hunter, known by the familiar title of "Whistling Joe;" and though in my boyish days I knew this burly, happy, healthy-looking hero well, from his being in the habit of calling at the school-house in his professional capacity, still I could never ascertain his surname; but it never gave me much trouble, since every one seemed perfectly satisfied to call and recognize him by the appropriate cognomen of "Whistling Joe the Ratcatcher."

A few months since, as I was walking from Ware to Hertford for the purpose of ascertaining something with regard to this well-known character, I met an old countryman and three young ones; when, for the purpose of drawing them into conversation, I asked my way to Hertford. "Go straight ahead," said the old man, "an' you'll run your nose bang agin it!" After such a polished piece of information, I inquired if ever he knew a man by the name of "Whistling Joe." This question had the effect of transforming him at once into a civilized being. "Oh yes, sir," said he, "I know'd him; he wa' a terrible chap! Do you know, sir, when I wa' a young man he gi'd me just what I wanted?"—"Indeed, and what was that, pray?" "Why, he gi'd me the soundest hiding I ever had, and broke two o' my ribs for insulting his dog. Still, what wa' better nor that, he never let me want, but looked to me every day till I was able to go to work again."—"But how did you insult his dog?" "Why you see, sir, I thought I was as good a man as he; so I kicked his dog."—"And for that he broke your ribs?" "Just so, sir; but, lor', talk o' fighting, why there wasn't a man round about as dared look at him in that way; he wa' the terror of the whole county. At the same time, sir, he wa' as gentle and kind-hearted as a gal. And as for dogs, traps, and keepers, he wa' afraid o' nothin'. He wa' a fine fellow; but, like all the rest o' the good uns, he's dead now!"

Here the old man gave his head a grave and melancholy shake, and finished his mournful sequel with a sigh, which made the young men look seriously down at their toes. After a minute's pause, I asked him if he could tell me what Joe's name was. "Why, 'Whistling Joe,' to be sure," was the reply. "Yes, yes," I returned, "that was his nickname; but what was his surname?" He said he never heard as ever he had any other name. I then inquired if he knew Joe's father. This question seemed a puzzler; for, after taking his hat off with one hand and for a time scratching his head with the other, he hesitatingly replied, "No; he couldn't say as how he ever did know Joe's father—indeed, he never heard as ever he had one."

With this piece of lucid information I thanked him, and walked on; yet felt perfectly satisfied that, according to the old-fashioned course of things, Joe must have had a father, and that that father must have had a name, and that that name, whatever it might be, Joe could claim as his birthright.

Here the reader will excuse me if, for a few moments, I intrude upon his patience by giving a slight description of the sermulot-hunter as he is, though to describe "Whistling Joe" and his exploits would be to paint a picture of nearly the whole of the fraternity; still, as I take up the subject on universal and not individual principle, I may be allowed to describe the genuine provincial sermulot-hunter as he is mostly found, and then continue the history of "Whistling Joe."

The provincial sermulot-hunter is a compound of three distinct characters—the gipsy, the gamekeeper, and the poacher. He possesses all the distant independence and dauntless bearing of the gipsy, with the knowledge of game, dogs, ferrets, &c., of the gamekeeper, and the quiet, midnight daring of the poacher. He looks upon all men alike, but humbles to none; and, being a man of secrets, he is of necessity a man of few words; and hence it is, from his mysterious bearing, coupled with so many marvellous anecdotes of his wonderful exploits, that a long-standing impression has been entertained that he deals in supernatural agency for the purpose of charming the sermulots from their hiding-places; and thus it is that he is looked upon with an eye of fear and suspicion by those who are simple enough to believe in such nonsense. By habit and custom he is very distant and retired in his manners, and seldom answers an inquisitive question with anything more than "yes" or "no;" that is, if he answers at all. But, if you expect him to tell the truth, you may find yourself most wofully mistaken, as he always keeps a rigid guard upon his tongue for the preservation of his secrets. As for Joe, it appears, he always answered every prying question by whistling a stave or two of "Walker, the Twopenny Postman," or any other popular tune; and hence his name of "Whistling Joe the Ratcatcher."

Drunk or sober, let no man molest one of these worthies. If he does, let him look out for the dogs; for, upon the least incitement, some three or four, or half a dozen, of them will bury their fangs in his legs, and thus avenge the insult offered to their master. Thus—asleep or awake—by night or by day—at home or abroad—does he live in unchequered security through the affection of his rough-muzzled equipage; the fidelity of which, few are found foolhardy enough to dispute, or put to a practical test. Nor does the first peer of the realm lay himself down with more conscious security than does the sermulot-hunter beneath a hedge on a sultry summer's day. Yet, without an exception, all that I have met with seem to labour under an impression that they live, not as it were by sufferance, or common consent, but in spite and open defiance. Hence their quiet-revengeful disposition; for, like gipsies, Jews, and brigands, they never forget an injury till it is fully avenged, though it may be for years to come. Thus it is that they have fallen into almost universal disrepute, which has tended much to the diminution of their numbers; for farmers unfortunately seem to think, with Hamlet, that it is better to "bear the ills they have than fly to others that they know not of." Thus, for years past, numbers have refused to employ them, while others pay them so insufficiently that the profession now is scarce worth following, and several have left it for other occupations. But it is a well-known fact that they make the best of gamekeepers and the worst of poachers. And it is a matter of greater question, whether any man is fully qualified for a gamekeeper till he is well skilled in the art of ratcatching; for if a full investigation were instituted, I believe, from their numbers and powers of increase, that rats or sermulots would prove to be the greatest enemies keepers have to contend with.

"Whistling Joe," at one period of his life, was employed as a gamekeeper, and, so far as daring and experience went, he was, doubtless, quite equal to his situation. But, at the same time, he had an unqualified notion of freedom and independence, and, like most uneducated men, he thought the only way he had of showing it was to insult those above him. Consequently, in one of these ignorant fits, he gave serious offence to his master, who, after paying him his wages, at once dismissed him from his service.

"I suppose, then, you mean it," said Joe. "I do," replied the gentleman. "Oh, very well," returned Joe; "but of course you don't expect any shooting next season." Then turning upon his heel, he left the house whistling.

The gentleman of course saw something very ominous in this remark; and, in despite of every precaution, he soon found it sensibly verified in the daily disappearance of his game; therefore, to save the remainder, he sent a polite invitation to Joe, which ended in his being reinstated, on the express condition that, for the future, he should treat his master with common civility.

Some years since the papers gave an account of one of Joe's drunken exploits. A man who bore Joe a deadly hatred for sundry lumps and bumps which he had received at his hands, happened to come into a public-house where Joe was drinking. He had with him a dog, to which Joe took a great fancy, and, forgetting all animosity, he wished to know what he should give him for it. The man was morose and sullen; but Joe would have no denial, and insisted upon his drinking with him. After drinking together for some time, they began to barter about the dog, when Joe offered him a guinea for it. "Yes," said the vindictive owner, "you shall have it for a guinea, if you will let me have a fire at you with a bullet at a hundred yards;" at the same time pointing at Joe's gun that was reared in the corner. "Agreed," said Joe; "here's the money, and there's the bullet. Now load the gun, and come on." The gun was quickly charged, and out they went into a neighbouring field, where Joe paced a hundred steps; then taking his station with his legs astride to steady his reeling carcass, and waving his hat above his head, he called upon the man to fire. The malicious scoundrel took a deliberate aim and fired, and so nearly did he hit his mark, that the ball ploughed up the ground as it passed through Joe's legs. "Now," roared Joe, "the dog's mine;" and so the matter ended.

At any time, when "Whistling Joe" fell short of money, or was hungry from an over-night's debauch, he would prepare some sermulots, and broil them on the gridiron for his dinner. This I am most credibly informed was a common occurrence in the taproom of a tavern, where, with buttered sermulots, a loaf, and a tankard of ale, he would enjoy himself with all the satisfaction that an alderman does his turtle, &c., at a city feed; and such was the apparent relish with which he ate them, that frequently the lookers-on have asked him for a taste; and on being supplied with a leg or shoulder, they have declared them to be very nice eating. "Nice eating," Joe would say, "to be sure they are. Why shouldn't they be? For what animals are cleaner or feed nicer than they do? The truth is, one-half the world are fools, an' don't know what's good." This denunciation was always followed by a satisfactory swig at the ale.

As to the underground rats in towns and cities, "Whistling Joe" was of opinion, that although they may injure the masonry, or rather brickwork, of the drains and sewers; still, if human life be of more value than bricks and mortar, let them, said he, preserve the one and repair the other. At the same time, let the authorities beware how they allow sewer-rats to be maltreated; for until permanent plans can be devised whereby the drains and sewers shall be continually cleansed by sudden flushes and rushes of water, they will find the rat to be that subterranean friend that saves them from periodical plagues, the result of deadly gases arising out of the putrefaction of animal and vegetable matter. But in the event of success in flushing, then comes the remedy which does away with the utility for rat-hunters, because the rats, being deprived of foothold, must of necessity be drowned or washed out with the current. Yet, in those districts where the current cannot or does not reach, let the inhabitants (though justified by every law in reason and nature in resorting to any and every means in their power to eject them from their premises) still bless God for his bounty in sending them so valuable, persevering, and salutary a benefactor.

"But I must tell you," said "Whistling Joe," "that there is already a great cause for the diminution of their numbers, and that is the now extensive use of pottery pipes and traps for drains. Were all the drains of London properly secured with traps or gratings, then would the rats be kept in the sewers, and the result would be that they themselves would keep their numbers in check, for this reason—most old male rats will, at any time, kill and devour the blind sucklings of their own species; while, on the other hand, the females, being securely shut out from the houses, would find very few places of security in which to litter and suckle their young. Like doe-rabbits and some other animals, when they find that the male has discovered their retreat, they will devour their young themselves rather than let him get at them. Thus, by this double species of infanticide, would sewer-rats keep their numbers within bounds, if by proper drainage they were kept within the sewers; and those who (from parsimony or carelessness) neglect their drains, to say nothing of the nauseous stench they endure, have the foundation of their premises dangerously drilled, and their establishments rendered mere breeding-grounds, or rat-preserves, to supply the underground avenues with scavengers. But in the country, where rats are of no earthly good, save to fertilize the soil with their carcasses, they are allowed every facility for increasing and multiplying, to the detriment of farmers and the nation's loss.

"Whistling Joe" has given his opinions on ratcatchers. Considering the great value of the services of these men, nothing could be more unjust, he conceived, than the mean consideration they receive at the hands of farmers; for if it be worth a London merchant's or hotel-keeper's while to give a ratcatcher six pounds a year to keep his premises clear, pray how much more shall it be worth a farmer's while, where his property on every side is exposed to the ravages of vermin?

Of all men you may wrong or insult, beware of the ratcatcher. He may not upbraid you or openly resent an injury, but look after your corn and cattle, for if his revenge be deep-seated and determined, he may inundate the one, and perhaps poison the other. Nor is it to be wondered at; for of all the ratcatchers I ever met, I never knew one that could read; and, for the most part, observed but two principles, namely, secrecy and silent vengeance. They live, as it were, in tranquil defiance. Their companions are but few, and their conversation limited. They are feared by one portion of society, and despised by the other; hence it is that they live a life, as it were, bordering on outlawry.

Speaking of the barn-owl, "Whistling Joe" maintained that of all birds in the feathered creation, the services of this bird were the most necessary and the most valuable to agriculturists, not only for their consumption of all kinds of mice, but rats also. "I have seen both a dog and a cat," said he, "refuse to kill a shrew; but the owl will swallow it with the utmost satisfaction. I have also seen an owl stoop and grasp a three-parts-grown rat across the neck and loins, and bear it away; and when I arrived at the nest, I found the rat torn to pieces, and half devoured by the young, which the old one was feeding."

"Whistling Joe's" account of himself and his ratcatching propensities was rather curious. "My history," said he, "is soon told. I was born in India, and of English parents. My father was a British officer, and sold his commission to come to England for the benefit of his health. My parents had not been here long before they died of the small-pox, and left me an orphan in the wide world without a relative to look to or care for me. I was turned out of doors by the landlord, and took to the high road to find a friend. I walked till I was foot-sore; and cold and hungry as I was I laid me down in a ditch by the road side to sleep. In the dead of the night I was awoke by a number of dogs sniffing around me, when Michael Finnacy—in honour be that name mentioned—helped me out, and after kindly inquiring my history, asked me if I would like to go with him, and be his son. I was but too glad to cling to any one who would feed me, and treat me kindly. He pulled a puppy out of his pocket, and gave it me; then setting me across his shoulders, with the ferret-bag over my arm, he bore me to his home; and hence I became a ratcatcher.

"My parents had been careful of my education; and being a clever reader for a child seven years old, and able to write, it afforded an endless gratification to my generous benefactor. Though I never lost sight of my learning, but always studied to improve it, still I had entirely forgotten my name; for time went on merrily, and from a continual habit I had of whistling, I was universally called 'Whistling Joe' the ratcatcher. But when I was eighteen, my benefactor fell ill, and, after a short confinement, died, and left me all his property and stock in trade. By this I was set up in life, and possessed of a decent sum of money.

"About this time the trials of John Black and the old squire commenced. Those trials ought never to be forgotten, though now they have sunk almost into a legend. They took place before the present generation. Suffice to say, I was engaged by the squire himself to clear his place of rats. One morning, while quietly seated in an outhouse behind some tubs, the squire, with Bush and his companion, two notorious poachers, entered. I heard the squire bribe them, and saw him give them fifty guineas each, besides promising them situations as gamekeepers, to perjure themselves by swearing an alibi, by which the squire was afterwards acquitted, and John Black entirely ruined. But scarcely had the bargain been settled, when I was discovered behind the tubs. The squire offered me the same terms; and because I refused to become a perjured accomplice, I was sentenced to twelve months' imprisonment for an attempted robbery. The magistrate was in the conspiracy; consequently it was useless attempting any defence; therefore all I told the squire was to beware of rats. After my release I was kindly received, being a favourite with all who knew me; for I always had an open hand and a dollar for a friend in distress. There was not a gamekeeper or ratcatcher in the country but what supplied me with vermin to carry out my design. I spared neither pains nor trouble to inundate him. I swarmed his preserves with wood-rats and other means of destruction, till he had not a head of game left. I made his lakes and ponds completely alive with water-rats. I drew them from all the farms round about into his; nor was there a rat-catcher in the district that would come to his assistance for love or money. Lastly, I filled his house and outhouses; and had he not taken fright, and sold his estate to go abroad, I would have kept on increasing them till they had eaten him alive in his bed. Nor do I repent of the act; though to this day I have ever been looked upon as using supernatural agency.

"But I must tell you that the chief means I used was trailing red-herrings, calves' tails, &c., with the exception of occasionally using what is known as '*the oil of rats*,' which is a most effectual thing. To obtain this oil, you must skin a sufficient number of rats, then take out their bowels, but leave in the heart, liver, and lights; then put them into an earthen jar, tie them over with a sheet of paper or bladder, and put them into a hot oven; there let them stand till they are quite dissolved, and the oil will float on the top; skim it off with a teaspoon, put it into a small phial, and preserve till you want it for use.

"When my kind patron Squire Wilson bought his estate, he sent for me to come and undo all that I had done. He knew John Black well, and all the circumstances connected with the trials, and also how I had served the late squire. He offered me a hundred guineas a year to be his game-keeper, which offer I refused, as I could not endure the idea of servitude. He then offered me that sum annually, if I would clear the estate, and restock the preserves with game, which offer I accepted, and did away with the necessity for a keeper; for there was not a poacher round about, that would not sooner give me a head of game than take one. As for Bush and his companion, they had been transported for highway robbery. Thus far I was secure; and as Squire Wilson was going on his travels for three years, he gave orders to supply me with everything I wanted. I had the thing entirely in my own hands, and being on good terms with all the keepers round about, I was plentifully supplied with leverets and eggs, both pheasants' and partridges'. When the leverets were too young, I suckled them under cats or rabbits, and as to the eggs, I hatched the partridges' under bantams, and the pheasants' under larger hens. When the squire returned, after three years' absence, I had perhaps the finest stock of game, for the size of the place, in all England. Gentlemen and gamekeepers used to come far and near to see it; and it was the admiration of all who came. I became the squire's confidential man for forty years; and when he died I retired, and have lived upon my means ever since."

In Part II., under the heads of "Trapping" and "Poisoning" of rats (Chapters XI. and XII) much useful information will be found, derived from "Whistling Joe's" long experience.

CHAPTER XIX.

MISCELLANEOUS ANECDOTES OF SERMULOTS AND SNAILS.

THE "Quarterly Review" informs us, that when sermulots get possession of any bird-breeding island, where the ship may have stopped to take in water, they invariably drive away the feathered inhabitants, by plundering their nests of both eggs and young. In this way were the puffins driven off Puffins' Island.

The puffin (observes Mr. Bewick) is a foot long, with scarcely any tail, and weighs about twelve ounces. But from the size, shape, substance, and sharpness of its bill, it is commonly called the coulterneb or knife-bill. It lives a great deal on various kinds of shell-fish, which it is enabled to do from the great powers of its bill and jaws, which enable it to crush the shell, and pluck out the fish. At the same time it is a most dangerous weapon of defence, especially to small animals, since the bird can cut them almost to pieces with a few chops or bites.

The female makes no nest, but deposits her egg on the bare mould, in a hole dug in the ground, or in those she may find ready-made by the rabbits, which she and her companion soon dislodge. They will also take shelter in these holes in stormy weather. They are very kind and attentive to their young, which they defend to the last against an enemy, and will suffer themselves to be killed or taken prisoners rather than desert them.

The bite of these birds is very severe; for one that was sent to him in a box, covered over with a net, made a snap at a poor man's fingers, and brought away the flesh as if it had been cut out with a knife.

They congregate in immense flocks on the cliffs and coasts of Britain and Ireland. The island already spoken of, which stands off the coast of Caernarvon, and which took the name of Puffins' Island, from the myriads of these birds that used to inhabit it, is now entirely deserted by them, on account of the ravages and onslaughts of the sermulots.

Now if sermulots will attack, and put to the rout, so formidable a bird as the puffin, with its great powers of defence, pray what chance is there for the young of game and poultry? But we will leave this for the consideration of farmers' wives, and those gentlemen who feel interested in the preservation of poultry and game.

The number of young ducks which the sermulots destroy in the Zoological Gardens, renders it necessary to surround the pools with a wire fencing, which half-way up has a pipe of wire-work, the large circle of which is not complete by several inches in the under-part; and the sermulot, unable to crawl along the concave roof, which stops his onward path, is compelled to return nonsuited.

The sermulots have been for a long time the pests of the gardens, being attracted by the quantities of food. The grating under one of the tiger's dens is eaten through by this nimble-toothed burglar, who makes as light of copper wire as of leaden pipes. Directly the new monkey-house was built, they took possession by gnawing through the flooring in every direction, for the sake of getting at poor Jacko's bread. Vigorous measures were resorted to, to defeat them; the floors were filled in with concrete, and the open roof was ceiled; but they quickly drilled their holes through the lath and plaster of the ceiling, which may be seen at this day. They burrowed in the old enclosure of the ground till it was quite rotten; and they still march about the den of the rhinoceros, and scamper over his horny hide when laid down to rest. It is only by hunting them with terriers that they can keep them under; and as many as fifty in a week are often killed, and their carcasses handed over to the eagles and vultures, who quickly transport them to their digestive regions, and then eagerly look out for more.

Not long ago, a gentleman living at Maidstone caught, by means of a net in a pond at Millgate, fifteen species of pike. One of them measured three feet five inches in length and twenty inches round, and weighed twenty pounds. When it was opened, they took from its inside another pike, measuring fifteen inches in length, and weighing a pound and a half; besides a great toad. Then, upon opening the extracted fish, they found a smaller pike, and a large sermulot. Thus the poor sermulot had been twice swallowed by these ravenous tyrants of the waters.

By the generality of mankind, the snail, as an article of food, is looked upon with the same degree of loathing as the detested sermulot; still we find our Gallic neighbours feed upon them with the same *goût* as they occasionally do upon the frog or the rat. In truth, snail-eating in Paris has become a kind of luxury. In a return of the statistics of Paris, it is stated that snails have become quite as fashionable an article of diet as they were in the days of ancient Rome. There were then in the city of Paris fifty-eight eating-houses, and one thousand two hundred private families, where snails were eaten daily, as a delicacy, by between eight and ten thousand individuals. The monthly cost for snails in Paris was estimated at half a million of francs. The market-price of the great vineyard snails was from two shillings to two and sixpence per hundred, while those from the hedges, woods, and forests, only fetched from one and sixpence to two shillings per hundred. The return then states that the proprietor of the snailery of Dijon realized seven thousand francs a year by the sale of his snails.

Now let us reduce this to something more easy of comprehension. In the first place, we will set down the snail-eaters at ten thousand; secondly, we will calculate the cost according to the standard of English money, and then we shall ascertain the expense, and the average number of snails each individual consumes daily. Now half a million of francs is rather better than £20,833 sterling. In the next place it costs each individual, on the average, about £2. 10*s.* per lunar month, that is, 10*s.* 6*d.* per week; and if we calculate the snails at two shillings per hundred, and allow sixpence per week for carriage, then, in the aggregate, the whole ten thousand consume five hundred per head weekly; that is, rather over seventy-one snails for each individual daily.

Here it must appear evident to every one, from the high price of these snails, that they are not the food of the poor, but the luxury of the rich; for what poor person could afford to lay out ten and sixpence a week for snails.

Now may we not fairly ask which any one would soonest have for dinner—a roasted monkey, a boiled sermulot with onion sauce, or seventy-one snails?

PART II.

TO THE FARMERS OF GREAT BRITAIN,
ON THE
FECUNDITY AND DEVASTATING CHARACTER
OF THE RAT.

CHAPTER I.

ON THE UNIVERSAL PREVALENCE AND DESTRUCTIVE HABITS OF THE RAT.

BEFORE I begin to describe or give the results of my calculations, as to the enormous powers of the rat for increasing and multiplying, and also the immense losses sustained through its all-consuming voracity, let me appeal to the serious consideration and attention of farmers in general; since the whole of my undertaking, from first to last, has been directed to their personal and individual interest. To them it is matter of vital importance, so far as in numerous cases it makes all the difference between their losses and profits—in a word, between their poverty and their well-doing; for herein will they begin to see how they have been struggling for years with millstones around their necks. But if they will lay aside all foolish prejudice, and read and reflect with the same kindly spirit in which I have written, I shall show them how, by a simple effort, to throw the burden from their shoulders, and lay the axe to the root of their misfortunes, and that, too, without any outlay or expense to themselves; nay more, if they will follow implicitly my plans and advice, they will not only put a period to their misfortunes, but, from the moment they carry them in execution, they shall, according to the extent of their farms and vermin, gather wealth daily; and by so doing, they will not only enrich themselves, but increase British prosperity and native independence.

In the first place, do not believe for one moment that I am going to dictate to the farmers, or lay down rules and plans for the better cultivation of the various soils and acres on their farms. Such an attempt on my part would be neither more nor less than the veriest impudence and presumption, since I at once confess that my knowledge of agricultural matters is of a most limited character. In the second place, should I, in the course of my observations, let fall any remark or remarks that may appear at all pointed or personal, I trust my readers will receive them with a kindly spirit, as my most anxious desire is to heal, not to wound, and it is quite impossible for any physician to cure his patient until he has made himself thoroughly acquainted with the nature and causes of the complaint.

I shall now commence by giving a few general statistics, collected at various times, and arranged under separate heads, according to their respective counties, countries, or localities.

The following important details were published by a gentleman from Shropshire, of whose veracity and local knowledge of the subject the author has no doubt.

The writer commences by stating a few facts as to their numbers on a farm of a given size, and the damage they occasion to both landlords and tenants, without any equivalent benefit to either; "for in this country," he says, "they are not turned to any account, as they are in France, where, first, they are used to breed maggots from, for catching frogs; secondly, their bones are boiled down to make size of; and, thirdly, their skins are used to make gloves with." He then proceeds to relate that at a barn and small homestead near a pool, but a considerable distance from the farm-house, principal farm-buildings, and rick-yards, a boy was employed at spare times to catch the rats by means of a wire snare at the end of a stick, bent and fastened down in the usual way with a wooden peg. This lad, at this small but distant homestead, in about four months, caught no less than 630 rats. "Now, is it more than fair," the writer says, "to suppose that those he did not kill, which were spread about the farm-house and principal farm-buildings, where all the grain was put into barns and stacks, should amount to as many more?" These added to the above, would make 1,260. But he is quite certain that if all those which were in holes in the hedge-rows at a distance from the buildings and other parts of the farm, living in the open weather on field produce, poultry, game, &c., were also added, the number would be doubled. But let us put them down at 1,260 rats to be provided for out of the produce of a farm of 280 acres, and we shall be quite satisfied of the enormous losses yearly sustained by the farmers of the country.

We are told of another farm of 400 acres, not two miles distant from the one just spoken of, where, when the ricks were taken in for thrashing at Christmas, and the barn was finally emptied, the number of rats killed was over 1,400. Numbers escaped up the drains, and into the rat and rabbit holes in the adjoining hedges and covers. If the quantity of those which so escaped and those that were on other parts of the farm be added to the above number, can they be fairly reckoned at much less than 2,000 rats? And this number the farmer himself declared they far exceeded.

At another property, on a farm of about 180 acres, and of stiff and unfavourable land, where there were very few rabbits, the quantity of rats was truly frightful. The squeaking and gibbering at nights, as they crossed the roads and scampered about in search of food, were quite incredible. The hedge-banks were beaten quite bare with tracks to the rat-holes, and the holes themselves were perfectly countless. Here the farmer employed a boy with six or eight traps, who caught five or six of a night during the winter months. But at the emptying of one barn, the farmer himself stated that they killed 800 more, which together made 1,340 rats; and the numbers that escaped were so great, that the total seemed scarcely diminished by what had been killed.

On a farm of 330 acres, the farmer, who said he looked sharply after the rats, admitted that, upon an average, he caught three rats every night the year through, which made 1,095 per year. And yet how many still remained uncaught the ricks too plainly showed.

In a great many places the potato, turnip, and other fields, after harvest, swarmed with them to an incredible extent.

The writer then states that he could detail almost endless

similar accounts, which he had received from various farmers, upon whom he could place the utmost reliance. In short, without multiplying instances, but taking the average quantity of these animals, according to the numbers actually killed on various farms, it is impossible to estimate the numbers in the district here noticed at less than from 1,000 to 1,200 rats to be provided for out of the farmer's stock, with the aid of eggs, poultry, game, &c., upon each farm of 300 acres.

Now let us pass from Shropshire to Middlesex, where a rat-catcher was employed to kill rats at twopence per head. In one barn he killed about 250; and the men on the following day killed more than 200. But to show that this work is not all profit, the rat-catcher on this occasion had two fine ferrets killed by the rats, which were worth a crown each. He said there must have been at least 2,000 rats upon the farm, and yet some of the farmers of this district wished to contract with this man to kill their rats at £2 a year, and to call every month, or thereabouts. This would be only 3*s.* 4*d.* for each call, and each call would occupy a day, sometimes more, if he did his duty.

The same man was employed by a farmer of Little Ealing, when at one killing he destroyed over 200, for which he received twopence per head; and upon another occasion he was employed by some gentlemen of Windsor to provide them with a quantity of rats to decide matches between their dogs. He went to a little rick in the corner of a field, and succeeded in catching over 150 alive. Of course many were killed, and many escaped. However, the gentlemen gave him sixpence a head for the live ones. He states that the stack stood only the height of his own head, and looked more like a haycock than a corn-stack.

But to run through this man's exploits in rat-catching would be an almost endless task, or at least too long for my present undertaking; therefore suffice it to say that he informs me that he and his partner are the principal rat-catchers within eight miles of Hanwell, making a circumference of about forty-eight miles, which comprises an area of about 144 square miles of country thickly populated with rats; and yet these men cannot live by their legal calling, but are compelled in the summer to resort to night-fishing, or anything else, to get an honest living.

Before we dismiss this man, it may be as well to mention one circumstance, which in itself is so disgraceful that it deserves publicity; yet, for charity's sake, I will withhold the author's name.

He once called upon a near neighbour, to see if he would give him a job, as his farm was literally swarming with rats. To this request he received the following answer: "They won't give me anything for my corn now, so let the rats eat it!"

I ask what right has such a man as this to be eternally grunting and growling for a reduction of rent, when at the same time he makes such a wilful sacrifice of his property? On the other hand, show me, if you can, a more heart-rending sight than to see a British yeoman pleading poverty at the rent-table? Such, I am sorry to say, is sometimes the case, over which there is no control, except by the plan I am now pursuing with regard to rats. But this I must say, that of all men in the wide world to deal with, give me an English gentleman; he is ever open to reason and justice, and such is his unqualified love of both, that upon a fair conviction he is every ready to make an honourable concession in behalf of either. Consequently, if a farmer is careful and industrious, and at the same time most scrupulous as to the cleanliness of his farm with regard to vermin, and yet, withal, cannot make ends meet, why then it is quite clear there is something radically wrong, and I am willing to believe that, upon a fair representation of his grievances to his landlord, he will, after investigation, cause such an amicable adjustment of the rent as will enable his tenant to pay his way, and live like a British farmer. But which of us would like to pay money out of our pockets to support another in laziness and luxury. Or which of us would feel disposed to submit to the losses occasioned by wilful and malicious destruction of his property?

Before I leave the county of Middlesex, I will mention the following circumstance:—

A farmer of Hanwell was telling me of a rat-match he had with a neighbour, in which, to use his own words, he backed his dog against the other's for a thigh of mutton and trimmings; and, in order to facilitate the match, and also to be present at the supper, a third farmer undertook to have one of his ricks unthatched, wherein he was certain there were plenty of rats. The whole party met at the appointed time, and the rick was taken down; when the dogs killed rat for rat, till the rick was completely removed; and, on each party counting their dead, they found that the dogs had killed 170 each. But my informant complained that his opponent took an advantage of him, for, on his leaving the ground, he had scarcely gone half-way across the field, when the other, who had seen some rats escape into a hole in a piece of old wall close by, went and removed a stone, and pulling out nine by their tails, let his dog kill them; thus winning the match by nine rats, to the great amusement of all present, except the loser, who by this time had returned to witness his defeat. The number of rats killed in this one rick amounted to 349.

I was told by a man who for seven years was in the employment of a farmer, that he had seen a sackful of rats taken out of one barn. His master always had from five to fifteen ricks standing; but every rick was built on the bare ground, with the exception of a little straw at the bottom. He regularly employed a rat-catcher, who used to call about every month; and, as the ricks were not built on staddles, he always found plenty of rats to kill. My informant can form no idea as to the numbers destroyed at each killing; but he says that at times they were immense. I mention this circumstance merely to show the folly of building ricks upon the ground, because of the great facility it affords for vermin; but I shall have some remarks to make presently upon that subject.

A few years ago appeared an account of a capture of rats, which took place in a stack of beans, when 216 were taken alive, and 96 killed. The bean-stack had been standing twelve months, and the delay in thrashing it had led to this great increase of rats.

The author of the "Book of the Farm," states that he witnessed the exploits of a man in the art of rat-killing by means of twenty-one small steel or gin-traps. But some days before setting to work he had to put the place in proper order, and then in the course of one day he killed no less than 385 rats, for which he received only a penny a head.

As a gentleman was superintending the removal of a rick of beans, he found that a large number of migratory rats had taken up their temporary abode therein. He immediately had that portion of the rick-barton completely surrounded with boards, except a small hole for them to get out at; and so well was his plan laid, that he and his dog succeeded in destroying upwards of 700. The dead carcasses of the vermin filled four large wheelbarrows.

Not long ago, about 1,490 of these destructive pests of the farmers were killed in some very old standing wheat-ricks. The quantity of grain destroyed by these vermin must have been very great, as they had long held undisputed possession of their comfortable quarters. At this time corn was fetching from 75*s.* to 85*s.* per quarter.

Some time ago I was told that a farmer, had six large corn-ricks standing in a field close to his house. They had been standing a long time. However, there being a considerable rise in the market at the time of which he spoke, he gave orders for the ricks to be taken down and thrashed; but the instant they set to work, two of them fell to the ground, being complete shells, and not a grain of corn in them; and the hearts of the rest were so completely eaten away, that it was currently reported in the neighbourhood that the whole six ricks did not yield anything near the quantity of grain that ought to have been produced from one.

I once knew a man who lived in 1835 with a gentleman who had been a military officer, but who had left the army and turned farmer. In the same year a gentleman and near neighbour sold his hounds, and went to Italy, and, as a matter of course, left but few servants to look after the house, &c. His gardener was to have the produce of the garden for his salary till his master returned. It was the spring of the year, and my friend being an amateur gardener, they were growing peas and beans against each other for tankards of ale, and used to visit one another's garden, to watch the growth of the crops.

When my friend called one morning, he found the poor

gardener in a state of perfect consternation and bewilderment; but he no sooner saw his neighbour than, without saying a word, he took hold of him and conducted him to the garden, to show him the cause of his grief, when, to his surprise, not a single plant was standing. When he was asked how such a disaster came about, the gardener took him to the kennel at the back of the house, in front of which there was a pond, and showed him the monsters that had torn up his plants for the sake of the seeds at the bottom. Our hero says that the kennel and all round the pond was swarming with rats; and so daring and audacious did they appear, that neither he nor the gardener had courage enough to disturb them, but sought security by quietly leaving them to themselves. He returned home, and informed his master of the circumstance, who seemed very much infuriated at the marauders. But, strange to say, the rats, finding nothing to eat at the dog-kennel, that very night paid him a visit, and quietly ensconced themselves in his bean-rick. When he arose in the morning he soon discovered their presence, and directly ran to acquaint his master, who (like a gallant and intrepid soldier) instantly despatched him with orders to fetch the rat-catcher, and also his dogs and ferrets. Away he went, and arrived just in time to catch the man as he was starting across the country in search of a job. He took him into the gig, with his dogs and ferrets, and returned with equal speed to the farm, where they found the general seemingly red-hot for action. "But stop, my friend," said the general, "let us first hold a council of war;" upon which he took them into the parlour, and brought out a bottle of brandy, and glasses. "There, my friends, be seated, and let us reason the matter coolly." They seated themselves round the breakfast-table, and after demolishing sundry wedges of cold ham, bread, butter, eggs, &c., and washing them down with copious cups of coffee, each of which, by the bye, had been slightly diluted with small drops of brandy, they then drew themselves up for grave deliberation. "Now, my friends," said the general, "the time has arrived when our most daring resolutions and best energies must be brought into action; therefore I wish to consult your superior judgments as to which of two plans will be the best to discomfit and defeat the enemy, and I call upon you to furnish me with your most serious and unfettered opinions. The first mode of attack I propose is to down with the stack, and kill the enemy as they fly ; the second is to undermine their fortification with gunpowder, and blow them and beans into the air together."

His two companions gave their judgments in favour of the first plan, as the latter would destroy all the beans. "As for that," said the general, "the beans could be picked up afterwards ; but as you have given your verdict in favour of the first mode of attack, I of course bow to your decision ; and now, gentlemen, let's draw out for action." The rat-catcher put his ferrets into the stack, two of which were soon killed, without dislodging any of the enemy. By this time the men arrived, and set to work to remove the stack, when out ran rats in all directions, numbers of which on every side were killed by the dogs or bludgeons. However, at the conclusion of the battle, the number of the slain nearly filled a tumbril or dung-cart.

Thus did the determination of the general not only save his beans and other articles from destruction by rats, but rendered an essential service to the whole of the surrounding district.

We will now pass into Suffolk, and notice some of the exploits of a rat-catcher, in that county. In twelve weeks only this celebrated rat-catcher brought to one public-house in Ipswich no less than 11,464 rats for the purpose of sport, which he sold at half-a-crown per dozen, and took from the town in that short space of time no less a sum than £254. He declared that another year of this sort of work would have made his fortune, and expressed his full conviction, that when the magisterial authorities put a stop to rat-pitting, they struck a severe blow at the prosperity of Ipswich and its surrounding country.

The author of a work entitled "London Labour and the London Poor," has given the results of his researches and calculations as to the number of rat-pits existing, and the rats destroyed in the metropolis in the course of the year. He says there are forty public rat-pits in London, the chief of which are held at public-houses, and that their average destruction of rats is about twenty per week, or 1,040 in each year, in public matches; which in the aggregate shows an annual destruction of 54,080 rats, besides those destroyed in private practice. He goes on to state that a dealer in live animals told him that there were several men who brought a few dozens, or even a single dozen, from the country—men who were not professional rat-catchers, but working in gardens or on farms, and who in their leisure time caught rats for the London market.

The same gentleman further states that from the best information he could obtain, there were no fewer than 2,000 rats killed weekly in public and private trials, or 104,000 yearly; that the public exhibitions are only periodical, while the private training of dogs in this respect goes on uninterruptedly. But all this is exclusive of the rats killed by the profession, where they are employed annually, or by the job, to destroy or keep the premises clear of these destructive creatures. It must be borne in mind that nearly the whole of these 104,000 are genuine farm-rats, many of which are brought to London by various gardeners' and farmers' men; and the remainder by a body of men who, in the garb of rat-catchers, prowl about the country within a few miles of town, but who will turn their hands to anything to earn a scanty subsistence. They have no fixed destination, though London is the grand market and centre of attraction, and thither they repair as often as they succeed in capturing a few rats, or what not, which can realize a scanty supply of money. The number of these men amounts to between seventy and eighty, and their average earnings are about fifteen shillings a week. As to the permanent rat-catchers of London, they amount to about twenty-five in number; but these men could not live by their profession only; consequently they mostly deal in dogs, ferrets, rabbits, song-birds, pigeons, &c.

The two principal rat-destroyers of London (Mr. Shaw and Mr. Sabin) inform me that they destroy between 8,000 and 9,000 each annually; that is, averaging 17,000 a year between them; and it is seldom that they have much less than from 100 to 1,000 rats each on hand for sale. One of them showed me and my son between 800 and 900 of these creatures, all at one view, in cages; and the other showed me over 2,000 at one time, and warranted every one to be a genuine barn-rat. It was Christmas-time, and every rat was destroyed in matches before the week was over. Their dead carcasses weighed twelve hundredweight. He says that he pays the countrymen, &c., over £200 a year for rats; and both of these rat-destroyers tell me that they cannot keep a hundred rats anyhow for less than ten shillings a week.

Here it may be as well to mention that when a match is made between two dogs to kill a certain number of rats each, or when a dog is matched to kill rats against time, one of the articles of agreement mostly is, that one and all shall be fair barn-rats; consequently sewer-rats are wholly excluded, except when a greenhorn makes a match with a sharper, wherein he backs his dog to kill a certain number of rats within a given time, and allows the sharper to find the rats unconditionally; in that case his dog is almost certain to lose, as the sharper will take care to provide none but the largest sewer-rats, which possess enormous powers and daring, and inflict most severe and dangerous wounds.

Now, I think it a fair conclusion to arrive at, that out of the 104,000 rats which are annually destroyed in London, 100,000 are genuine country rats, while the remaining 4,000 will amply account for all the sewer, house, and wharf rats that may be destroyed in dog and ferret matches.

There is a person residing in London who is perhaps one of the most extensive, as well as one of the most honest, dog-dealers in England. It may seem curious to mention the latter quality; but, generally speaking, this class of worthies, like Yorkshire horse-dealers, wear their honour so loosely that they ought to walk hand-in-hand together. This man informs me that some little time since, he sold a bull-dog to a customer of his some few miles from town, where he carried on the double occupation of farmer and brewer. On the morning after the dog's arrival, his new master gave orders to his men to clear a certain barn, for the purpose of catching rats, to give him a trial. The men set to work right merrily, as there was to be some sport; and without hesitation seized the rats by the tails as they came across them. The smaller ones were dashed against the walls, while the larger ones were as carefully popped into an upright tub. When all was ready, the dog was brought, and put into the tub with the rats; and, as might be expected, a desperate battle ensued, as there was no escape for either. But before ten minutes had elapsed all the rats were dead, and, upon being turned out and counted, they numbered seventy.

Now, would it be more than fair to set down those that were dashed against the wall (which comprised all the smaller and half-grown ones) at less than double that number, since none but the largest were selected for the tub. Then add to these at least seventy more that escaped beneath the flooring, &c., which would make in all 280 rats in that one barn.

In order to show that England is not the only country or island that is infested with these unwelcome creatures, I will adduce a few details as to their numbers and doings in other portions of the world. "Generations had passed away," says the "Scotsman," "without seeing a rat on the small Island of Tarinsay, on the west coast of Harris; but an innumerable swarm of these annoying and destructive vermin have of late made their appearance on the island, and in despite of the tacksman's endeavours to extirpate them, they seem to be multiplying faster and faster. Indeed, they seem to be increasing so fast that they threaten to overrun the entire island, and take violent possession of it. They are supposed to have come from the island of Soay, which lies at the distance of about three miles from Tarinsay, and into which the Earl of Dunmore, some time ago, ordered a number of rabbits to be sent. Soon after this, the rats, which were very numerous in the island of Soay, seemed to take offence at the rabbits, and completely disappeared, having removed in a body to the neighbouring island of Tarinsay."

The most singlar part of this account is, that rats should take offence at rabbits; when, at the same time, they are among their most determined enemies. But such is the peculiarity of rats, that, if they take offence, or have a whim in their heads, they gather all together, and away they go in spite of every obstacle.

Some travellers have given accounts as to the number of rats on some of the islands in the Pacific, and as to how they migrate from island to island, or rather are supposed to migrate, on rafts and wrecks. Among the rest a gentleman has published his travels, entitled, "Four Years in the Pacific," wherein he relates his visit to Juan Fernandez. This is the island so celebrated in England as being the place whereon Robinson Crusoe is represented to have been shipwrecked, and where he first met with his man Friday. Here the author met with a family of settlers in the greatest poverty and distress, and who doubtless had been led there through reading the "Life and Adventures of Robinson Crusoe," as depicted by Defoe. The land was very rich and fertile, and vegetation spread forth in rank profusion. It appears it was only necessary to plant corn in the ground, and it would rise up in abundance, so far as the ground was concerned. They were living upon a small farm, which was allowed to run to waste; and when the author asked them why they did not cultivate the ground, they replied, that it was no use, for the rats destroyed all the grain, and hence arose their privations.

That rats do migrate from island to island, and that for miles across the sea, there is no doubt; but how they migrate, that is another question. For my own part I can easily believe that when a ship is in harbour they in the night-time swim in a body from the shore to the vessel, and having clambered on board, take their berths without paying for their passage; and that when the ship anchors at another station, either at home or abroad, they, under the cover of night, swim in a body from the ship to the shore, and then walk away without saying either "Good bye" or "Thank you."

It often happens that ships will become so completely overrun with them that the owners are compelled, the first fair opportunity, to make the vessel as air-tight as possible, and burn charcoal and brimstone in the hold, or between decks, to suffocate them; but, in that case, the ship for some time after will stink most filthily from the effects of their putrid carcasses, though vast numbers may be found dead round the charcoal pans or dishes.

Here I may mention one curious circumstance with regard to ship-rats. It has often been matter of surprise that they do not sink the vessels by boring holes so as to admit the water. It is a well-known fact, that, however they may gnaw their way through the interior of the ship, they never pass through the sides, though in numberless instances they have eaten half-way, which has led to strong discussions as to the reasoning powers of the rat; but Mr. Jesse, I believe, gives the most correct reason; he believes that they gnaw through the sides till the wood becomes distasteful from the effects of the salt water with which it is constantly saturated, and which renders it so disgusting and offensive to their tongues, that they desist from further operations—not from the power of reason which would apprehend danger, but from the saline deposits towards the outer surface of the wood, which is both loathsome and nauseous to their palates, and hence the safety of ships.

If ships in general had on board a number of such passengers as had the good steam-ship *Ohio*, when she arrived at San Francisco, not only the interior of ships, but the cargoes also, would be greatly preserved from the ravages of rats. The *Ohio* had on board over a hundred cats, for which they paid at the rate of sixpence each, and at California sold them for from ten to twenty dollars per cat. This may give some idea of the vast quantities of rats in the gold-regions.

When the steam frigate *Terrible*, under the command of Captain Hope, returned from the Mediterranean, and was lying at Woolwich, she was found to be swarming with rats, notwithstanding the numbers that had been daily killed by the officers and crew with sticks while running about the cabins and other parts of the vessel. A rat-catcher was had on board, who, in the course of three days, caught over 400 alive, and the ship was afterwards fumigated, to suffocate the remainder.

The mischief done by rats on board ship is not confined to the victualling department only, but anything composed of vellum or parchment, whether in the shape of wills, deeds, or directions, equally suffers from their predatious propensities. The West-Indian steamer *Prince*, on one occasion, brought back the European mails, in consequence of the rats having eaten off all the parchment labels. It is curious that rats will not touch leather bags of a tan colour, but are extremely fond of parchment. Some time since a will in England was required in Demerara, and, after a great deal of trouble and expense, it was obtained and sent out; but upon the mail's arriving at Demerara, the rats had devoured the will, leaving nothing but the seal behind.

The following circumstance will give some idea of their consumption of food on board ship, and also their numbers. On the return of the man-of-war ship *Valiant* from the Havanna, its rats had increased to such an alarming degree, that they destroyed above a hundredweight of biscuit daily, besides other food that came in their way. The ship was at length smoked between decks, in order to suffocate them, and six hampers were filled daily for some time with rats that had been thus killed. In this case these creatures had been daily devouring more food than 112 sailors; or over a ton and a half of biscuits monthly; to say nothing of other articles.

In the Isle of France rats have been found in such prodigious swarms, that the place was completely abandoned by the Dutch on account of their overwhelming numbers. The author of "A Million of Facts," speaks very positively on this point, by stating most emphatically that the numbers of rats drove the Dutch from the Isle of France. In some of the houses they were so numerous that 30,000 have been known to be killed in a single year. They have subterraneous hoards both of corn and fruit, and frequently climb trees, to devour the young birds. At sunset they may be seen running about in all directions, and in a single night commit such devastations, that a French traveller says he has seen a whole field of Indian corn in which they have not left a single ear remaining.

The public journals, some thirty years since, were filled with the most alarming and distressing accounts from Van Dieman's Land. The vermin had come down in such immense hordes upon the cultivated grounds, and were increasing so fast, that they threatened destruction to the entire settlements. But I am most credibly informed, that two ship-loads of rat-killing dogs were sent out from England to the settlers, which had the effect of saving them from all the horrors of famine.

With respect to rats in India, I heard of a lady who some years ago had married a military officer, and left this country for India. The lady was delighted with the country and the social circle into which she was introduced.

There was, however, one appalling drawback, and that was the enormous size, numbers, and impudence of the hideous rats with which the place was infested; and such was their daring, that neither the chamber, drawing-room, nor indeed any part of the establishment, was free from their perambulations. If she met one on the stairs, it would not turn back, but come boldly forward, and pass her by with silent contempt. At dinner-time they would walk into the dining-room with all the confidence of a pet dog or cat, and scramble over her feet after the crumbs that fell, while the older ones would clamber up the cloth on to the table, and help themselves to what they pleased, and then depart, without noticing either her or her husband, or indeed any one else who might by chance be dining with them. But, as a matter of course, numerous quarrels and fights continually took place through others striving to rob the thieves of their stolen treasures. However, at the time the lady was writing, she had become pretty well reconciled. Thus it is, that use will make us familiar with almost anything. But what made the matter worse, there was no possibility of getting rid of these vermin; for to kill them would be an endless task, and to poison them would only create some putrefactive disease in so hot a climate, from the dead carcasses that would be lying about in every hole and corner of the premises; so that to live near would be impossible. Indeed her husband knew but one plan to get rid of them, and that was to set fire to the house, and burn it to the ground. This, certainly, would be a most effectual plan, but, at the same time, a most expensive one.

From the number of facts already adduced in the course of this introductory chapter, the farmers of Great Britain may be perfectly satisfied that their numbers of rats, and the losses and annoyances they sustain through them, are by no means singular either at home or abroad. Let it be borne in mind that there are no annual or quarterly returns of rats, as there ought to be, and would be, were they wolves, though infinitely less destructive; but being pronounced vermin, no notice is taken of them. Consequently I have been driven to the necessity of availing myself of private as well as authentic sources, and also to note the circumstances as they appear recorded in our various journals, for the purpose of establishing the fact, that rats are not only a scourge to farmers, merchants, and others, but are starvers of the poor, and a hinderance to individual and national prosperity, in whatever part of the world they accumulate ; and that no cultivated portion of the universe is free from their devastating and ruinous effects. Were their numbers and depredations fully enumerated and recorded, I believe the narrative would reach to the North Pole and back again, and then leave an appendage which, when unfolded, would more than lap the circumference of the globe. For their numbers and devastations, in every part of the known world, are so multitudinous and overwhelming, that they are perfectly beyond all enumeration ; and a person might just as reasonably set to work to count the stars in the firmament, as to give anything like a correct account of the numbers and doings of rats.

I shall now proceed to arrange my subject under three principal divisions :—First, the FECUNDITY of rats ; second, their DEVASTATING POWERS ; and, lastly, the best means for their LOCAL EXTIRPATION.

CHAPTER II.

FECUNDITY OF RATS.

I SHALL here commence by giving the results of my calculations as to the prolific nature of rats, since their powers for increasing and multiplying are not, perhaps, to be surpassed by any other order of quadrupeds. In this respect I have been assisted by the best possible information, both from naturalists of the highest standing, and also from living individuals of long and practical experience.

Baron Cuvier and most other naturalists state that rats have from twelve to eighteen young ones at a litter. A celebrated rat-catcher declares that he once took from a dead rat no less than twenty-one young ones, and speaks of two other cases wherein the same number was produced, and another tells me of one case he heard of, wherein twenty-three were found. Nevertheless, I am satisfied from personal experience, that the average number ranges between eight and fourteen at a litter ; and that these large numbers are matters but of rare occurrence.

In the next place, I am informed that rats will have six, seven, and even eight nests of young in the year, and this for three or four years running ; that the young will breed at three months old ; and that there are considerably more females than males.

In order that I may not leave the smallest room for doubt or dispute, I propose laying down my calculations at somewhere about one-half. In the first place, I will say four litters in the year, beginning and ending with a litter ; thus making thirteen litters in three years. Secondly, to have eight young ones at a litter, half male and half female ; and, thirdly, the young ones to have a litter at six months old. At this calculation I will take one pair ; and at the end of three years what do you suppose will be the amount of living rats ? why, no less a number than 651,050 ; that is, including the father, mother, children, grandchildren, great-grandchildren, &c.

The enormous number of 651,050 living rats springing from one pair in three years, appears a thing almost incredible. Yet it is an arithmetical fact. Nevertheless, for the satisfaction of those who may feel curious in the matter, and have an inclination to investigate for themselves, I will give the round numbers of young, each littering, as they transpire. Therefore, suppose we commence with Lady-day 1856, and finish on Lady-day 1859, then we shall find the following to be the result :—

First litter, 8 ; second litter, 8.

In the third littering, the first young ones being then six months old, the four females, with their mother, will have a litter each, which, together, will number 40. Thus will they fall in to the finish :—

Fourth littering, 72 ; fifth, 232 ; sixth, 520 ; seventh, 1,448 ; eighth, 3,528 ; ninth, 9,320 ; tenth, 23,432 ; eleventh, 60,712 ; twelfth, 154,440 ; thirteenth, 397,288. Now add all these together, and the number will be 651,050.

But to make the matter easier to cast up, I will give the figures :—

In the first place, put down the two old ones ..	2
Lady-day, 1856	
Midsummer-day, 1856	8
Michaelmas-day, 1856.—The first litter are six months' old	40
Christmas-day, 1856	72
Lady-day, 1857	232
Midsummer-day, 1857	520
Michaelmas-day, 1857	1,448
Christmas-day, 1857	3,528
Lady-day, 1858	9,320
Midsummer-day, 1858	23,432
Michaelmas-day, 1858	60,712
Christmas-day, 1858	154,440
Lady-day, 1859	397,288
	651,050

Wonderful as this calculation may appear, yet what will be more wonderful still is, that presently I shall be able to prove that they would eat and waste more food than would suffice to feed over 65,000 human beings. Or, to render the thing easier of comprehension, let us suppose them to have but one litter more—making fourteen litterings in three years and three months—and then we shall find the numbers to be 1,015,048, which would eat and destroy more grain than would supply the whole of the British army with one pound nine ounces of bread each, day by day, even though they numbered 101,504 men.

CHAPTER III.

DEVASTATING POWERS OF RATS, AND THEIR ENORMOUS CONSUMPTION OF GRAIN.

At a late meeting of farmers, among whom the devastating qualities of the rat formed the subject of discussion, it was generally agreed, in reply to the author's questions, that, with eating and wasting, they would destroy half a pint of grain every twenty-four hours; and that six rats would eat, day by day, as much as an ordinary man.

Some few years ago, there was an account in the newspapers, to the effect, that an experiment had been tried with rats, as to the amount of food they would consume, and the results published were, that six rats would eat as much as a man, but that eight would eat more than a man: therefore, to place the matter beyond all doubt, I will set down ten rats to eat as much as an ordinary man; and I sincerely wish that every man, woman, and child throughout the human family could secure, day by day, as much food each as ten rats can dispose of; and then we should have no more bad harvests, famines, and their appalling consequences.

Now, allow me to give you a slight estimate as to your losses. Suppose there are five hundred and twelve wine-glasses or half-gills, at strike measure, in a bushel; if five hundred rats eat only a wine-glass, or half-gill of corn each, every twenty-four hours, and you allow twelve glasses only for waste, then they dispose of a bushel of corn per day. We must also bear in mind, that it is not only what they fairly eat, but what they waste also, which must be considered; and if any of you have watched them feed, as I have done, you will bear witness to this fact, namely, that the number of grains they eat, in comparison to those they waste, is regulated according to the quality of the corn. It is true that when corn is sent to the mill, the good, bad, and indifferent are all equally ground into flour; but, as the rat grinds his own corn, he selects that which is most easy and palatable for the purpose; and as he grinds it grain by grain he naturally selects the best, where there is plenty. The result is, I have at times seen them bite through and let fall four, six, and even eight grains out of ten; and as they will keep on till they have their fill, the consequence is, that the worse the corn, the greater is the waste.

I have just counted the number of grains in half a gill of the best wheat that money could purchase; it is now, at the time of writing, fetching seventy-two shillings per quarter. There are twelve hundred and twenty grains, and out of these there are four hundred and twenty which are small, shrivelled, or unsound; consequently there is more than one-third of it which would be wasted; that is to say, if five hundred rats had a bushel of this corn to eat, instead of three half pints, they would waste over two gallons and a half; or, to make it plainer, they would, in eight days, get rid of a quarter, or eight bushels, out of which they would waste over two bushels and five gallons. Now, if this be the waste of this superior and expensive grain, pray what must be the waste among poor and inferior corn?

The usual custom of rats is to average six months' residence out of the twelve in the barns, ricks, granaries, &c.; say from Michaelmas to Lady-day. During the other six months they mostly resort to the fields, hedges, ditches, woods, and the water-side, and feed upon crops, eggs, birds, and fish of all kinds that come within their reach, or indeed anything else that chance may throw in their way. However, we will say nothing about what they consume during the summer months; but calculate, to a certain extent, the losses they occasion in the farms and farmsteads during the winter months.

I shall now state the amount of losses according to the wine-glass standard, and the number of the rats you may calculate you have upon your respective farms. A party of farmers with whom I was lately in conversation, gave different estimates of the number of these vermin on their respective farms. One estimated them at 250; consequently his losses amounted to half a bushel per day, or eleven quarters three bushels and a peck in the half-year. Another gentleman estimated his rats at 500; consequently his losses amounted to a bushel per day, or twenty-two quarters six bushels and a half in the six months. The chairman of the meeting estimated his rats at 750; consequently his losses amounted to one bushel and a half per day, or thirty-four quarters one bushel and three pecks in the half-year. Three of the gentlemen estimated their rats at 1000 each, and, as a matter of course, their losses would be proportionately greater; that is to say, a thousand rats, eating only a wine-glass of corn per day each, and allowing in the same time only twenty-four glasses for waste among a thousand feeding, they will, in six months, or 182 days and a half, consume five bushels over nine loads of British corn.

Let us next calculate the loss in the shape of bread. I

do not mean London bakers' bread, but bread pure and unadulterated. One bushel of corn, I believe, will make a bushel of flour, and a bushel of flour will make sixteen four-pound, or thirty-two two-pound loaves. Well, then, as your losses are estimated at two bushels per day, that is equal to supplying sixty-four men with a two-pound loaf each, every day for six months, or thirty-two men the year round, and so in proportion with every other person.

Here let me again remind you, that this calculation is only on the supposition that a rat will not destroy more than a wine-glass of corn, at strike measure, per day. But if I take your unanimous opinion that a rat will eat and waste half a pint of grain per day, why then your losses are four times the greater, and instead of one bushel per day you lose four.

If we calculate that ten rats will eat, day by day, as much as a man, why then you are keeping equal to a hundred men in idleness, for six months in the year, or fifty men the year round; at the same time saying nothing about what rats may rob you of during the summer months, say from Lady-day to Michaelmas.

In fact, the whole of your taxes put together, poor-rates and all, cost not so much as it does to feed your rats.

And now I beg to suggest a plan which, if carried out, would, I believe, prove that rats daily destroy more grain, &c., than would suffice to provide for all the poor and indigent in the British empire. This plan is, to raise a universal warfare against the whole of the rat-tribe in every farming district in England, and to compel the various churchwardens or overseers to take down the dates and numbers destroyed in their respective parishes; so that their accounts might be fairly contrasted with the returns of the Poor Law Commissioners during the same period.

I challenge the test, and feel no apprehension for the result; since I am strongly of opinion, that what with black rats and brown rats, they would be proved to have run away with infinitely more than one-fifth of the farmers' corn and the poor man's loaf.

In the first place let me calculate the losses sustained by rats on a farm near Upton-on-Severn. The paper states that they amounted to 1,490. Let us suppose that only ten made their escape, that would make the number 1,500, which, at the wine-glass standard, would destroy three quarters of grain every eight days; that is sixty-eight quarters three bushels and a half in six months. Now these were killed when corn was, and had been, fetching from seventy-five to eighty-five shillings per quarter; we will divide the difference, and set it down at eighty shillings per quarter. Now what did it cost the farmer to keep his rats for six months, at this calculation? No less a sum than £273 7*s.*

"Yes," it may be said, "but corn does not always fetch eighty shillings per quarter." Agreed; and now, to meet this point fairly, I will reduce the price to one-half, say forty shillings per quarter, instead of eighty, and then what would it cost to keep these rats for six months?—why, £136 17*s.* 6*d.*

Now let me put it in another shape, and suppose that in the broad day the Government of the country called upon the same man to give away this amount of corn in the shape of bread to the poor of his district, that would be a two-pound loaf each every day for six months to ninety-six poor people, or forty-eight poor people the year round. Do you think he would do it? or would he rather reach down his sword and blunderbuss and strike for a revolution? Yet, at the same time, he loses the same amount of corn in the dark by rats, and says little or nothing about it. But let me put his losses in the third shape; that is, if ten rats will eat and waste as much as an ordinary man can eat, then he was keeping equal to 150 men for six months in the year, or seventy-five individuals the year round, in laziness.

I will now put your losses into something like shape. According to the same calculations, the farmer who estimates his rats at only 250, loses only eleven quarters three bushels and a peck in the half-year, which, at forty shillings per quarter, amounts only to £22 16*s.* 3*d.*, which, to say nothing of the odd peck and bushels, is equal to giving sixteen men a two-pound loaf each per day for six months, or eight men a loaf each per day the year round. But if we allow ten rats to eat as much as an ordinary man, why then he is keeping equal to twenty-five men for six months out of the twelve, or twelve men and a boy in laziness the year round. The farmer who estimates his rats at 500, loses just double the amount, namely, £45 12*s.* 6*d.*, which is equal to giving thirty-two men a two-pound loaf each per day for six months, or a loaf each per day to sixteen men the year round; or at ten rats per man, he is keeping equal to fifty men for six months in the year, or twenty-five men the year round in laziness. He who estimates his rats at 750, sustains a loss of a bushel and a half per day, or thirty-four quarters one bushel and three pecks in the half-year, which, at only forty shillings per quarter, would fetch £68 8*s.* 9*d.*, which, to say nothing of the odd measure, would supply forty-eight men with a two-pound loaf each per day for six months, or twenty-four men with the same amount the whole year; but at ten rats per man it is equal to keeping seventy-five men in laziness for six months, or thirty-seven men and a boy the year round. Those farmers who estimate their number of rats at 1,000, will be sorry to learn that at the wine-glass standard they each lose two bushels of corn per day, or forty-five quarters and five bushels in the half-year, which, at forty shillings per quarter, would realize £91 5*s.*, or, if the corn were distributed in two-pound loaves, it would supply sixty-four individuals daily for six months, or thirty-two the whole year through; or if disposed of according to the standard of ten rats to a man, it would supply a hundred lazy men for six months, or fifty for the entire year.

Having arrived thus far, let us now bear in mind three distinct considerations. The first is, that these calculations are made at the low price of forty shillings per quarter for corn. But should corn be at sixty shillings, then instead of £91 5*s.* you would lose £136 17*s.* 6*d.*; or should corn be, as of late years it often has been, eighty shillings per quarter, then your losses would be £182 10*s.*

The next consideration is, that we set down rats to eat and waste only a wine-glass, or half a gill each, in twenty-four hours; but should they destroy a gill, then under any circumstances your losses would be double. And the third consideration is, that we have made no calculation whatever as to what they destroy during the other six months, in the shape of grain, roots, eggs, pigeons, poultry, &c.

I shall conclude the present chapter by noticing the peculiarities of rats and mice in a rick; and here I would ask you how many mice it would take to eat as much as a rat? Some say six, some seven, and others eight. But to make the matter easy of calculation, suppose we put down ten mice to eat as much as a rat. But rats, it is said, eat the mice. Very true; and so do cats eat the rats, when they can catch 'em! But as cats are too large to get into rat-holes, they don't often kill rats; and as rats are too large to get into mouse-holes, the consequence is, they don't often eat mice. The truth is, that in a rick they each keep their respective apartments. The rats, for the convenience of water, inhabit the lower part of the rick, and eat upwards; while the mice, to get out of their way, and also to quench their thirst with the dews and rain on the thatch, inhabit the upper part, and eat downwards; and what with rats eating upwards, and mice eating downwards, they, in a few months, make a very nice rick of it.

CHAPTER IV.

VERMIN-KILLERS AND RAT-MATCHES.

SUPPOSE, for the sake of illustration, we make an aggregate of all the British farms, and divide the rats and mice equally among them, according to their number of acres, and then what think you would be the average number upon each farm of, say, three hundred acres? Of course those that have their thousand or thousands would make up for those that have but hundreds, and supply also the model farms, where they may have but few. This is independent of the young they might have. I will again give the Salopian gentleman's estimate as to the average number of rats in his district. He says that he could give almost endless accounts which he has received from various farmers, upon whom he could place the utmost reliance. In short, without multiplying instances, but taking the average quantity of these animals according to the numbers actually killed on various farms, it is impossible to estimate the numbers in those localities at less than from 1,000 to 1,200 rats, to be provided for out of the farmer's stock, with the aid of eggs, poultry, game, &c., upon each farm of 300 acres. Here we have the average number of rats in Shropshire.

But as other counties may not have so many, suppose we set down the smallest number as the average of farm-vermin in England, namely 250, with the addition of twice the number of mice, which would be 500; and if we average ten mice to a rat, that will be equal to 300 rats.

I find, according to the census of 1851, that, without Ireland, there are in Great Britain and the islands in the British Seas, 303,720 farmers, and only 2,256 vermin-killers; that is, not one vermin-killer to every 134 farmers. At the same time we must bear in mind, that, to say nothing of the shipping, all the towns, villages, and hamlets, with all the warehouses, factories, and storehouses of all kinds, throughout Great Britain and her Channel Islands, would be left without a single rat-catcher among them. Then how perfectly unequal to the work is the number of rat-catchers; but we will leave them till by-and-by, and go on with the calculations.

This is rather an alarming state of things, still it is only averaging the farms at 300 acres each; and supposing them to have but one rat, or ten mice, to an acre. In truth, it does not signify what size the farms may be, so long as we average one rat, or ten mice, to each acre; but, for the sake of calculation, we will lump them together, and then suppose them to be divided into equal portions of three hundred acres; and what do you suppose would be the number of vermin upon the 303,720 farms? An amount equivalent to 91,116,000 rats! which, at the wine-glass standard, would consume 182,232 bushels of corn daily, or 4,157,167 quarters and four bushels in the half-year, namely, 182 days and a half; and which would supply 5,831,424 people with a two-pound loaf each daily for six months, or 2,915,712 people daily with a two-pound loaf each the year round. Or, at ten rats per man, it would support 4,555,800 people for twelve months; that is, about twice the population of London and its suburbs.

Let us now see what it would cost in money to feed these creatures. Suppose we take grain at the price it is while I am writing. Corn is 75*s.*, barley 43*s.*, and oats 32*s.* per quarter. Let us lump them together, say one-third wheat, one-third barley, and one-third oats, and then they would average 50*s.* per quarter; then averaging wheat, barley, and oats at 50*s.* per quarter, it would cost £10,392,918 10*s.* to feed these rats for six months, with a wine-glass of corn each per day; and all this is without making any calculation for the young they would have during the same period.

I now propose laying down a few abstract questions for the due consideration of our worthy magistracy, as to the propriety of their interference in rat-matching, &c. The first point I would wish to draw their attention to, is the case of the rat-catcher of Witnesham, in Suffolk, who brought, in twenty-one weeks, to one public-house in Ipswich, no less a number than 11,465 rats, which he sold at two shillings and sixpence per dozen. This man, in this short space of time, took from this public-house the sum of £254; and all these were genuine barn-rats, which were brought for the purpose of training dogs, matches, &c.

The first matter for consideration is simply this,—would this man have taken the trouble to have either caught or killed that number of rats in the same time, but for the consideration of the money he could make by them? Then he declares, that another year of the same sort of work would have made his fortune, and expresses his full conviction, that when the authorities stopped rat-pitting, they struck a severe blow at the prosperity of Ipswich and the surrounding district. But the matter for consideration is the amount of human sustenance these animals were daily destroying. If we allow that ten rats can eat as much as a man, why then they consumed more food than would support, day by day, 1,140 individuals. This rat-catcher laments the interference of the authorities, because they laid the axe to the root of his prosperity: "for," says he, "another year of the same work would have made my fortune;" thereby implying that there were still myriads of rats remaining at his command in the district.

Let us now suppose that our friend had been left alone for a year. What then would have been the result, if we judge from those he had already caught, which, with the exception of one rat in twenty-one weeks, was at the rate of 546 weekly, or 28,392 in twelve months; and which would consume daily more food than 2,839 human beings.

I shall now lay before you the complaint of a landlord of

a tavern. His house was situated about a stone's throw from the police-station. Here he was landlord for seven years, and, though considered something of a sporting house, still, during that period, he declared that not one case of disturbance ever went from his tavern to the station. Nor was he himself ever charged with any misdemeanour in connection with his house. Yet because this man had rat-matches in his establishment, and trained dogs in the art of rat-killing, he was persecuted by an eldery functionary, till he was compelled to leave his house.

Let us calculate but a portion of the benefit this man's occupation conferred upon mankind, and then you can form your own opinions as to the wisdom and justice of his persecutor. But, to prevent too much repetition, I will couple his doings with those of another rat-catcher.

Mr. Charles Dickens, in his "Household Words," has designated these two men the "Napoleons" of the Rat-wars. These men destroy between eight and nine thousand each, yearly; so we will set that down as 17,000 between them; and these 17,000 rats would consume, the year through, as much food as would supply 1,700 men with sustenance for the same time. But this is without any consideration for the young they would produce during that period, which, if we calculate one-half to be females, namely 8,500, and to have five litters in a year (that is, beginning and ending with a litter), and to have eight young ones at a birth, and the young to breed at six months old, then within the twelve months they would produce a family of 3,060,000 strong. Then add to these the old ones, and you have a family of rats numbering 3,077,000; which, at the wine-glass standard, would eat 6,154 bushels of grain daily; and which grain would supply 196,928 men with a two-pound loaf each per day; or, at ten rats per man, they would eat food sufficient to supply 307,700 people daily.

Before proceeding further, let us calculate the amount of money it would take to feed these animals for twelve months. One thing is certain, that, unless driven by extreme necessity, they will never eat either beans or peas. The reasons are obvious: in the first place, they are too hard for mastication; and in the second, they are not delicate enough for their palates. Neither will they eat either oats or barley while corn is to be had; the truth is, they will leave all other grain for wheat, and will, at all times, leave bad for good; and as they are excellent judges, they will spare neither time, trouble, nor perseverance to get at the best. As to their being found sometimes in large numbers in bean-ricks, that is to be accounted for by their being very open and loose, and thereby affording them a ready shelter in their migrations. To them it appears to matter little whether it be a bean-stack or faggot-stack, so long as it affords them shelter for the day, or till they can work their way into better quarters. However, to place the cost on a fair foundation, suppose we mix equal portions of wheat, oats, and barley together, and average them at fifty shillings per quarter, then we shall find that to purchase grain enough to supply these rats for one year, at a wine-glass each daily, and at fifty shillings per quarter, it would cost £701,940 12s.

Now, had not these men's occupations rendered the nation a service, by not only destroying 17,000 rats, but cutting off their entire posterity, whereby all the corn, &c., has been saved which otherwise would have been destroyed?

Here let it be most distinctly and most emphatically understood, that I do not come forward as the champion of rat-killers or rat-matches. My object is too serious to think of sport; but from my knowledge of, and my investigations into the doings of rats, and the serious consequences they entail upon the country, it appears to me that some of our magistracy have not had the opportunity of giving the matter that deep consideration which it merits, before giving their judgment; and hence it is that I would draw your attention to the following calculations, in order that you may come to a conclusion as to the propriety of stopping the destruction of this class of vermin, even though it be in the shape of rat-pitting.

To say it is a matter of no importance as to what decision you or the legislature may come to upon the subject, would be a fallacy, since, I am sorry to say, there are too many honest parents who, having to work hard for a large family, know too well the difference in the baker's bill between seven shillings and fourteen per week for bread and flour, and that, too, without any countervailing benefit to either landlord or farmer. Hence I call upon you to think, and that deeply, before you decide upon the wisdom of throwing the law's protective cloak over the lives of vermin; for if you shield the rat—grotesque as it may appear—British justice will demand fair play, and compel you to protect the lives of every other class of vermin that afflict and annoy the human family!

I will not give my own account of the murders of rats destroyed yearly in London in matches and private practice, but prefer giving and substantiating, as near as may be, the calculations of Mr. Henry Mayhew. After the deepest investigation, he came to the conclusion that there were no fewer than 104,000 killed annually in London in public matches and private training; that is at the rate of 2,000 per week. At the same time bear in mind that nearly the whole of these are professedly country-rats, all of which are caught on the farms, &c., in the outskirts of London.

Now, if a few individuals from the outskirts can supply the London market with 2,000 rats per week the year through, pray what must be the state of the entire country? Still it may be said that some of them are sewer-rats. That, doubtless, is true; but, from the best information I could arrive at, and for reasons I have already stated, I am satisfied that not four out of a hundred are caught in the sewers. But to make it satisfactory, and at the same time more easy of calculation, I will set them down at 4,000 in the year, thus leaving 100,000 to be brought from the outskirts; or, more plainly speaking, common barn-rats.

Let us now suppose rat-matching, &c., in London to be suspended for one year only, and calculate the amount of food they would consume during that period, and then we shall see a portion of the benefit conferred upon mankind by their destruction. Thus 100,000 rats, at 10 rats per man, would eat as much food as 10,000 people. Then, if we calculate them at the wine-glass standard, it would take 200 bushels, or 25 quarters per day, to supply them with a wine-glass of grain each, at strike measure; and which, in the year, would amount to 9,125 quarters. Or if we calculate it in the shape of bread—supposing two pounds of flour to make only two pounds of bread—it would supply 6,400 poor people with a two-pound loaf each daily, the year round.

Moreover, the land necessary to grow the corn, say good land, yielding a fair average crop, namely, eight combs, or four quarters, to the acre, would amount to 2,281¼ acres, which would employ eleven farmers—ten at 200 acres each. and the eleventh at 281¼; and all this to keep the rats that are destroyed in London in one year in dog-training and rat-pitting.

We have not reckoned what the grain would cost. Suppose we calculate as before, oats, barley, and wheat together, and set it down at 50*s*. per quarter. Then we shall find, that to give these rats a wine-glass of grain each daily, it would cost £22,812 10*s*. for the year.

But bear in mind that all this is without any consideration of the young produced during the same period. Therefore, if we calculate them as before, namely, one-half females, to have five litters in the year, beginning and ending with a litter, and eight young ones at a birth, and the young to breed at six months old, then what would be the amount of living rats to be supported out of the farmer's corn and the poor man's loaf in the outskirts of London? No less a number than 9,000,000, which, at ten rats per man, would eat, day by day, the year round, as much food as would supply 900,000 people for the same time. That is nearly one-third the population of London and its suburbs, or nine times the number of the British army, if we estimate it at 100,000 strong.

Then, if we reckon them according to the wine-glass standard, what would be the amount of grain necessary to feed them for one day? Why, 18,000 bushels, or 821,250 quarters per year, which, if made into bread, would supply 576,000 human beings with a two-pound loaf each daily for the same period. The ground necessary to grow the grain, at eight combs, or four quarters, to the acre, would be 205,312½ acres; and this would employ 410 farmers at 500 acres each, and 312½ for another. And what amount of money would it take to purchase this mixed grain, at 50*s*. per quarter? No less a sum than £2,053,125—all of which has been most happily and most effectually saved through the destruction of the parents in the rat-pits, &c., of London.

The matter now for consideration is, as to whether

rat-pitting has not rendered humanity a service by exterminating so many of these destructive creatures. If so, then, is it not most desirable, or would it not be a most salutary step, if rat-pits were established in every town and village in the British dominions? It is useless to talk of the police of nature; for they, like the professed rat-catchers, have proved wholly unequal to the duties required of them; or to assert, that the craving of rats for each other's lives has proved a remedy for the evil; for if stoats, weasels, hawks, owls, rat-catchers, &c., have done their duty, or if rats are ever cruelly and wantonly destroying their own species, pray how comes it that they are found not only by the hundreds and thousands, but by the tens of thousands and hundreds of thousands, in a community (which the rats of Montfaucon and in the sewers of Paris prove to demonstration), while, at the same time, their numbers are spread over every temperate region of the known world? The subject admits of no argument, because the overwhelming numbers of living facts annihilate every species of speculative theory. But of the inefficiency of the various animals of prey to keep their numbers in check, I shall speak presently.

And now I will close this part of my subject by drawing the attention of our worthy magistrates to three other points, and then leave the matter in their hands for more mature consideration. The first point is, as to how far, and how often, during the past twelve months, has the public peace of London been disturbed by ratting? second, how far and in what degree have the public morals been debased by seeing or participating in the same? and, lastly, how many convictions have taken place in our courts of justice during the same period through dog-training and rat-pitting?

CHAPTER V.

THE SEWER-RATS AND RAT-CATCHERS OF LONDON.

I AM not an advocate for the indiscriminate slaughter of rats; for I am one of those who believe that nothing, however repulsive in appearance or noxious its habits, has been created but for some wise purpose. Hence I maintain that when rats are in the sewers, fulfilling the duties for which they were created, they are rendering mankind a most essential service in devouring the refuse of animal and vegetable matter which would otherwise go to putrefaction and breed a multiplicity of diseases, or rather an incessant plague. Under such circumstances, therefore, for any man to go into their haunts, and make them prisoners for the purposes of gain, or to kill them where they are, is neither more nor less than animal-murder. But, on the other hand, when they burst their boundaries, and intrude on the territories of man, to the danger of his children, his castle, and his properties, then he is not only justified in repelling them, as he would any other ravenous beast, but in killing them also, in order to save himself, his family, and his properties from their further inroads and devastations.

It may be asked, if the rats in the sewers were left unmolested, would not the number of deaths among them (which of necessity must be very numerous) produce a deleterious effect similar to the animal and vegetable refuse already spoken of? My reply would be, "Certainly not, for the best of reasons—because rats always eat their dead as well as the aged and infirm. Therein they prove not only a remedy for the evil, but a social, healthy, and orderly community."

However, let me leave this, and proceed with the accounts of Jumper and Jem, the rat-catchers of the sewers. They, with Jem's man Jack, are the only rat-catchers I can hear of in the sewers of London; Jumper's apprentice, to use his own words, having long since "cut it."

The London newspapers lately gave some account of a man who was enjoying his own mode of existence as much as anybody else, though he had chosen a career replete with dreariness and filth. In an assault case, heard before the magistrates at the Clerkenwell Court, Jumper appeared as a witness. He said that he was a rat-catcher by profession, and caught rats in the sewers for those gentlemen who keep sporting dogs. When he had an order, he went forth with a bull's-eye lantern, a strong wire cage, and a short rake; then, entering the main sewers at the foot of Blackfriars Bridge, he waded his way, "waist-deep," in muck and filth of the description which the closets and sinks of London furnished. With the rake he disturbed the creatures in their hiding-places, which, becoming dazzled with the light, fell an easy prey, and were forthwith caged.

It appears, however, that it is not to brown rats only that Jumper devotes his time and industry. There are gold and silver rats, in the shape of purses, watches, rings, and plate, which are of far more importance to him than brown rats. On one occasion he found a silk purse filled with gold and silver coin; on another, a gold watch and seals, he is constantly picking up silver spoons, rings, and other articles of value which may be accidentally dropped or thrown down the drains through carelessness, or pitched down the gratings in the streets by swell-mobsmen, prostitutes, or minor pickpockets when pursued.

Jumper's underground journeys extend for miles. He has been under Newgate, and along Cheapside, to the Mansion House, and has traversed from Holborn to Islington, closely inspecting all the various stinking tributaries which fall into the main sewers of the mighty metropolis. It is declared that he knows more about the sewerage of London and its condition than any other living man; and that, upon the strength of such qualification, he would make an excellent chairman of the Board of Commissioners sitting in Greek Street.

Some time ago Jumper took an apprentice for a month, on liking, prior to being bound to the profession. His name was Harris; he had been bred to the horse-slaughtering business; and having gone through such a course of preparation, he might be supposed to have lost the sense of smell and delicacy of stomach sufficiently to enable him to enter upon his new occupation. After a month's trial, however, he gave it up for a bad job. "I can stand a tidy lot," said he, "but I can't stand that 'ere;" so, to make things agreeable, they had a pot and parted; and thereby, Jumper and Jem, and Jem's man Jack, were left sole sovereigns of the sewers.

But they, like most other eminent men, bitterly complain of the jealousy and persecution they experience from meddling people, who, envious of their reputation, take pleasure in annoying them. A lord mayor, upon one occasion, disputed Jumper's right, and threatened to put him in prison for trespass; and upon another occasion, a police constable took Jem and Jack into custody, for being found unlawfully in the city sewers, and for having a key wherewith they could unlock the various iron trap-doors that cover in the side entrances to the sewers. But they were both dismissed, on their assurance that they did not injure the walls, and that the commissioners and their men never interfered with them.

So far, so good; but as to how far commissioners are justified in allowing these men in the sewers, or the lord mayor and the policemen in keeping them out, we shall presently see. In order to show the danger in its true light, let us suppose a conversation to take place between these men; not for the purpose of bringing disgrace upon them, but to expose, in an easy and familiar manner, the danger that exists, from allowing, not only free passage to the sewers, but persons to possess keys whereby they may at any hour of night or day, pass in and out at pleasure, and fasten the traps behind them.

It appears that after the dismissal of the charge just detailed, Jem and Jack left Guildhall, and made the best of their way to the Custom-house, to meet Jumper, who was then in the sewers, catching rats for Jem; he having the job of supplying two gentlemen with three dozen for a private match, wherein one had backed his dog to kill them in five minutes. This gentleman was a novice in such matches, and as the term "barn-rats" was not specified in the articles of agreement, Jem, as a matter of course, was left to do as he pleased as to what rats they should be. However, the gentlemen (unknown to one another) gave Jem a sovereign each, to make the wager, as each thought, "safe" for himself. The master of the dog was very anxious the rats should be as small as possible; and, on the other hand, his opponent wished them to be as large as possible. Now here was a dilemma, which to any one else but Jem would have been perplexing, but not so with him, for he had his own way of acting what he called "honourable to all parties;" that is to say, he served those best who paid him best. Therefore, before leaving, he secretly asked the backer of the rats what he was to do, as the other gentleman had given him a sovereign to get the rats small. "Do," said the gentleman, at the same time putting another sovereign into his hand; "there now, get some of the largest sewer-rats you can find." "All right, Sir," said Jem, and away he went, resolving, that as one gentleman had given him one sovereign, and the other two, there should be two dozen of the largest sewer-rats he could find, and one dozen of small, or rather barn-rats, and that was what he called "acting honourable to all parties."

However, upon arriving at the Custom-house he gave Jack a sovereign, to get boiled beef and pease-pudding enough for three, with plenty of vegetables, and two gallons of the best ale, in order to have, as he said, "a jolly blow-out over beating the Peeler." He then dived into the sewers after Jumper, while Jack went about his business.

Suffice it to say, that Jem had not been long under ground before he and Jumper were seated on their rat-cages, and settling the legality of the point in dispute, namely, their sovereign right to perambulate the sewers at pleasure. "I told the magistrate," said Jem, "that I was a master rat-catcher, an' that Jack was my man. 'Your worship,' says I, 'I've been at this kind of work for years, an' was never interrupted before. I catch rats for sporting gentlemen both at the east and west end of London; but Jack's only been nine months in the profession, and pretty clever he is, your worship, considering. However, your worship, we never injure the walls; an' when we come across the commissioners' men underground, neither they nor the commissioners ever interfere with us. Now, your worship, what I want to know is this,—what right had this man, with a shiny hat and tin buttons, to take us and the rats into custody, when we were coming quietly out of the trap, and interfering with nobody.' 'Truly,' said the magistrate ; 'and as neither the commissioners nor their men ever interfere with you, neither shall I ;' and so he burked the Peeler."

"Bravo !" said Jumper. "Now you see, Jem, this only proves what I have said a thousand times. If a man has any business to do, there's nothing like having a real out-an'-out lady or gentleman to deal with. 'Tis only necessary to make the thing clear to 'em, an' its all right. But as to your half-an'-half-bred varmints, they're a curse to society, an' would starve an' destroy everything around 'em. Now what could that skinny-looking, cupboard-hunting cur want interfering with you, when you were going quietly about your business ? It's ignerence, Jem, its all ignerence."

"But," says Jem, "what's to prevent our swell-mobsmen and house-breakers from getting keys of the sewers the same as ourselves, an' turn them into a regular fence, or at midnight kicking up a jolly fight to draw off the Peelers, an' when they're gone off to the station-house, those above ground to give the signal, while those beneath open up the trap, then crack any house they please ; an' if any one comes, crack them over the head with a jemmy, or let loose a barking-iron at 'em, pop them an' the plate down the trap together, then shut up shop ; an' who's to know what has become of 'em ? Why it would bother the Foresters themselves to find 'em out—an' as for the corpse, they could shove that on one side for eight-an'-forty hours, when the rats would pick it to the bone ; an' then they could double it up like a bundle o' sticks, and stow it away in the drain of some old house that's to let."

"Well now, Jem, I never thought of that before. Don't you see, Jumper, Ben of Horsington an' Appleton Bill call'd on me early this morning ; they're in a peck o' trouble. When they were honest rat-catchers, they were two as nice fellows as ever lived ; but as soon as they took to poaching they were spoiled, an' thought of nothing but desperation. However, they've just had three months at the mill, an' what did they do the night before last but crack the magistrate's house, an' bolt with the plate ; an' now, you see, they want me to lend 'em a hand, an' stow it away down here. But I told 'em, point blank, I'd have nothing to do with it ; 'an' as for Jumper,' I said, 'I know he wont, for he always dresses like a gentleman on Sundays, an' goes to church with a watch in his pocket. Therefore,' says I, 'Jumper an' I have always earned an honest living in the sewers, an' so we mean to remain.'" "That was right, Jem. But still, if these sort o' fellows once take possession of the sewers, then we shall have Sir Richard Mayne down here on his hands an' knees poking his nose into every hole an' corner, and establishing an underground police to prowl the sewers night an' day; an' if he does, then, Jumper, its all up with us." "I hope it won't come to that," said Jumper. "Nor I, either, for that matter. But still there's no end to the depredations such fellows could commit. Now look you here, Jumper, how easy could they wrench off the traps, an' let the vermin into all the houses, both new an' old. There are plenty o' these dog-stealing chaps would do it, if 'twere only to keep up the trade of rat-catching, &c. An' for the matter o' that, there are plenty of old drains, where, with very little trouble, they could get into the premises themselves." "That's very true," remarked Jumper. "An' then again, when there's a great poison trial, or what not, on at the Old Bailey, what's to prevent the prisoner's pals from undermining the place with gunpowder? Why, there are plenty o' these fellows who bet wagers on such life an' death cases as would give 'em thousands to do it rather than lose; an' when the trial's going against 'em, blow up the place, judge, jury, an' all the whole blessed lot together. Talk o' Guy Faux, or the storming of Sebastopol, why they wouldn't be a patch on it." "Really, Jem, its very alarming." "I know it is," said Jem. "But hark—Jack's coming with the dinner. Don't say anything about it before him, poor fellow, cos he'll only take to fretting."

Talk of Belshazzar's Feast, or of royal and imperial entertainments, or of the barons of old, or City gorges, or turtle or whitebait dinners, or of a feed with the Mutual Communication Society, at the Star and Garter, Richmond, with a view of the country from the two-floor back—pray what think you of a blow-out of boiled beef and pease-pudding in the City sewers, with bricks for chairs, a cage full of rats for a table, and the place illumined, not with chandeliers, but three bull's-eye lanterns? Those who have never experienced this delicate pleasure know nothing of the peaceful delights of it. At least, so say Jem, Jack, and Jumper.

"Is that you, Jack?" "All right, Jem," was the answer "But I had to leave half-a-crown on the can an' plates." "But where's the grub?" inquired Jumper. "The pape busted," said Jack, "so I turn'd the meat an' pudden int my hat, an' tucked it under my arm. As for the carrots an taters, I shoved them into my pocket, along o' th' plates But look; here's a jolly great she-rat I caught." "Let i go," said Jem; "poor thing, it's heavy with young. You should never hurt the shes. They're the best friends we have. Its all very well to catch the boars, when we happen to get a job, cos we might as well sell 'em as let 'em fight an kill each other. Don't hurt it, poor thing. Let it go gently 'twill find its own road back. There, that's it. Don't you see, Jack, if it were not for rats keeping the sewers sweet an clean by eating up the rubbish an' stuff, we couldn't put our heads into 'em. Ther'd be no end to diseases, an' fevers, an such-like things. Lor bless you, the people of London would die like rotten sheep." "Well, Jem, I only brought it to show you what a fat un it was." "All right, Jack. Now let's have the dinner, for I'm most uncommon hungry How do you feel, Jumper?" "Rather peckish," said Mr Jumper. "But stop, Jack, let's set the cage firm first, or the animals may upset it. There, now it makes a famous table." "Here, Jem, just hold the hat while I get a plate out o' my pocket—that's it. Now if I put the plate on the top, an' turn the hat upside down, then we shall have all the gravy, an' they gave us a famous lot o' mustard with it."

It is needless for me to give an exact account of the sayings and doings of these subterranean worthies while at dinner. But I will give a brief outline, and you can imagine the rest better than I can describe it.

When Jack had emptied his pockets on the table, they were forced to be quick, and put the vegetables on the plates, because the rats began pulling them through the wires into the cage.

"Now, my boy," said Jem, "first give your hat a welt against the wall to knock out the gravy and mustard; then plunge your hand into the mire there, an' grope about for a brick to sit on."

That done they took their seats, and set to work in right good earnest to devour all before them. One lot after the other disappeared, and were as punctually washed down with copious draughts of ale, till at last they were brought to an indescribable stand-still. Jack performed the office of waiter, who, after scraping the remnants of vegetables, &c., together, squeezed them through the wires on the rats, crammed the plates into his pocket, and then brought forward the pipes and tobacco. Jem, as a matter of course, having paid for the entertainment, was unanimously chosen to the chair. After the pipes were charged, and the three gentlemen well-nigh smothered in smoke, Mr. Jumper broke silence by volunteering to sing the "Rat-catcher's Daughter," which he executed in a style well worthy of the place, subject, and the occasion. That being finished, the chairman called upon him for a sentiment—"Certainly," said Mr. Jumper; "Here's to our Queen, God bless her, an' may she never enjoy her dinner worse than we have, or have worse subjects than us three." This sentiment gave even more satisfaction than the song, since it was considered a confirmation of their loyalty, and was knocked down accordingly, or rather clapped down, because the knocking on the cage made the rats nearly jump out of their skins. After that, the Canadian boat-song, "Row brothers, row," was sung in full force by the whole of the company. But suffice it to say, that, after sundry other songs and sentiments, they shouldered their cages, and reeled out of the sewers, because the ale was out, and then made the best of their way to the neighbouring pot-house to melt the remainder of the sovereign. The rat-match was not to come off till late in the evening.

Now that we are fairly out of the City sewers, let me conclude the subject of sewer-rats and rat-matches by referring our worthy magistrates, &c., to my accounts and statistics of the sewer-rats of Paris; and, by a careful perusal of the matter, I believe they will arrive at something like a conclusion —first, as to the wisdom of allowing strangers the liberty of going in and out of the sewers at pleasure, and to possess keys for their convenience; secondly, as to the justice of allowing them to catch and destroy the rats in the sewers while they are rendering mankind and the nation at large such an essential service; and lastly, as to the propriety of giving such offenders six months at the tread-mill for trespass and animal-murder. Here I most respectfully leave the matter in their hands for more mature consideration.

CHAPTER VI.

SUMMING-UP JUDGMENT AGAINST THE RAT.

HAVING now dismissed the magistrates, let us bring Master Rat to judgment upon the real merits and demerits of his doings and position with regard to mankind.

That it was a wise ordination of Providence which brought this animal to our shores at a time when our towns and cities began to increase in size and density, no thinking man can deny; for there is no doubt, but, from his persevering and boring propensities, and his all-devouring stomach, he renders mankind a most essential service, by keeping the drains and sewers in densely populated districts in a state of comparative health. Nature has provided him with every necessary quality both in form and inclination for this peculiar office; and for man to molest or destroy him in his legitimate province, or in the field of his salutary labours, is, in the eye of common justice, not only barbarous ingratitude, but downright animal-murder, and, in my humble opinion, ought to be checked by every legitimate means. But, on the other hand, when rats leave their territories, and intrude on those of man, to the destruction of his peace and property, then does the subject wear a totally different aspect. For it can be proved, by a world of witnesses, that he not only undermines the very foundations of our castles and out-buildings, but destroys our furniture and apparel, eats our wills and deeds, purloins our watches and jewellery, consumes our grain, destroys our poultry and game of all kinds, ransacks our dovecots, our gardens, our orchards, slaughterhouses, fish-ponds, larders, dairies, storehouses, &c., and perpetrates a host of other outrages both on sea and land; sometimes murdering our children in their beds, and devouring our departed friends and kindred. Then, so far from his being man's friend, he becomes his most daring and rapacious enemy. I say most daring and rapacious; for after summing up his various doings, show me, in the vast creation, one animal that is so universally destructive in its habits and propensities as the rat.

Therefore, after having given him a fair and impartial trial, the only just conclusions we can arrive at are, that in the drains and sewers of populous towns and cities he is of infinite value, so far as he renders mankind a most essential service in keeping those localities in a state of comparative sweetness; and in these localities he ought to be protected. But in every other situation in life, whether on sea or land, he is the most daring and destructive little brute in the whole animal creation, without any countervailing benefit, either in the shape of food or raiment (at least as regards England), and ought to be persecuted by every means in our power; for, if self-preservation be the first law of nature and the foundation-stone of every other law, then are we justified in carrying on not only a national but a universal war of extermination against the whole tribe of wanderers, not for the purposes of sport or vengeance, but for the common protection of ourselves and the human family.

CHAPTER VII.

RAT-CATCHERS, AND THE BEST MEANS FOR THE LOCAL EXTIRPATION OF THE RAT.

Now that we have tried and condemned Master Rat, let us come to the means of his local extirpation. The principal agents to be employed are ferrets, dogs, traps, nets, and poisons; but in no case is the last to be used where either of the other four can be used with effect. These will prevent all danger or accident, and a vast amount of animal-suffering.

My suggestions for carrying out this war are two; namely, to employ a skilful rat-catcher, or become your own rat-catchers; but in the event of your deciding upon employing a professional man, let me give you some instructions.

"As for rat-catchers," says a well-known author, "just find me an honest one, and I will forfeit my name; for I would as soon admit a colony of rats themselves to my house as one of these gentry; though I have been much amused by learning some of the slight tricks of trade from one of these representatives of roguery and unblushing effrontery."

This is indeed too sweeping and far too severe against a body of men whom the world at large so much injure by denying them a fair name, fair occupation, or even fair pay for their most valuable services, when most skilfully and most honourably executed. It brings to our minds the ancient philosopher, Diogenes, who, in broad daylight, walked through the streets of Athens with a lighted candle. Both friends and foes felt perfectly bewildered, as he gazed vacantly in their faces, and then passed silently on. The astonished multitude consulted together as to what could be the cause of this strange phenomenon, and resolved upon questioning him as to what he was seeking; and his reply was, "An honest man." So it would appear that honest men were very scarce even in those early days. But as to the present time, if our author had a candle that would reach from here to the moon, and made a candlestick of the London Monument, it is a matter of great doubt whether it would not burn out in the socket ere he could fill his hat with anything better than lawyers and Yorkshire horsedealers, whose honour and honesty, of course, no one can doubt. I am afraid it too often happens, when we cannot get our ends out of a man, that we stamp him with dishonour; and as far as my little experience will carry me, I have found Christians of nearly every denomination infuse too much of the Jewish maxim into their daily practices, which says, "Get money, honestly if you can, but get money."

I am satisfied that the genuine rat-catcher is a man who has never been duly appreciated; consequently his interests have never been duly respected, especially by those who, from the nature of his calling, ought to esteem him as their best friend and protector—I mean farmers. Thus has he often been driven to acts which, under more honourable treatment and a just consideration of his rights, he would spurn to do.

It is certain that every man ought to be able to live by his legitimate occupation; but if men deprive him of his just rights, why then, common necessity oft compels him to take to that which is less legal and less desirable. Therefore let the rat-catcher be fairly employed and justly paid; then he will refrain from poaching, &c., and be as honest as any other man. Be sure of this, that whatever you pay him, within reason, will be infinitely more than a thousand per cent. profit to you. Therefore earn this man's respect and good-will by kind treatment and generous pay. It will be to your best interest to do so. Remember that a willing, confidential servant will run far and wide, and do more, much more than he is paid for, to serve a kind and generous master; while, on the other hand, a dissatisfied, unwilling man will think his money hard earned by laziness, and will scarce thank you when you pay him.

Now, when the rat-catcher calls for his salary, see you do not find some insulting excuse for not paying him his fair demand. Do not tell him that you have no rats now, consequently have no more call for his services; because, if you do, and he is a man who well understands his profession, he will take care that before six months are past you shall have plenty; and if you be a dilatory farmer, and create this man's revenge, he can eat you nearly out of house and home in twelve months. He can bring them far and near to your estate, and lure the rats from all the farms round about on to yours; so that in six months you will be swarming alive; and at the end of twelve months you will scarcely have a grain of corn left to feed either them or yourselves.

In speaking of the rat-catcher's imaginary powers of bewitching rats, allow me to present the following anecdote.

A well-known rat-catcher of Mid-Lothian was, during the last long war with the French, caught by a press-gang, and taken on board one of the king's ships. At first he feigned madness; but finding that would not do, he betook himself quietly to work. However, it appears that the ship was overrun with rats; and one day he happened to overhear the captain and one of his mates consulting together as to the best means of getting rid of them. "What's all this about rats, when I'm on board," said the rat-catcher. "You on board," cried the captain; "and what can you do?" —"Do," said he, "why I can make them cut their own throats!" "The deuce you can!" exclaimed the captain; "then I'd like to see you do it!"

Away went our hero below, and got the blade of a razor; then rubbing it over with something out of a small bottle, came on deck, and stuck it upright in a plank; that done, he went down into the steward's room, and poured some of the stuff on the soles of his shoes; then, after walking about the place, he came up to where the razor was, and there pulled off his shoes; then went and stood apart, where the captain and the rest were waiting to see the result. Presently the rats smelt the stuff below, and following it, made their way on deck to the razor, when the eagerness of those behind pushed the foremost against the blade, and cut them. Like some of our friends here, the superstition of the captain overcame his judgment. He roared out to them to lower the boat instantly, and take that wizard on shore, for he wouldn't sail with him. "Take him on shore directly, I say, or he'll sink the ship." And so the rat-catcher was set at liberty once more, to follow his old trade of rat-charming.

And now, you are doubtless anxious to know what was in the bottle. But, first, I must tell you, there are four kinds of witches commonly used for charming rats; namely, a red herring, some old rags, a small bottle of the oil of aniseed, and a calf's tail. These are the witches! and the method of using them is simply thus:—If you prefer the red herring, all you have to do, is to tie it by the tail with a piece of string; but be careful to handle it as little as possible. Then, after dark, when all is quiet, just trail it on the ground round the barn or rick where the rats are, and then strike off, trailing it all the way to the place where you wish them to go, and there leave it; or, if you know the place where they drink, it is only necessary to trail it across their path to the place you would have them go, and that will have the same effect. The consequence is, when the rats come out to drink at the nearest pond, ditch, or river, they will catch the scent of the herring; off they will go, nosing it all the way like hounds; and when there, it is a hundred to one they do not go back, but quietly take up their abode where they are. This method of drawing them mostly proves successful; but either of the others I think better, which is, to fasten a string to the old rags or calf's tail; then pour some of the oil of aniseed upon them, and trail them the same as the herring to the place where you wish the rats should go. The rags should be old ones that have been exposed to the air; but, in any case, you should handle them as little as possible, because, if the rats smell the odour of your hands, they will not go, but run away. In the third place, if you would like to follow the other plan, all you have to do is, to pour some of the oil of aniseed on the soles of your boots, and take very short steps as you pass over the ground, the same as in trailing; but, in that case, when you arrive at the place, you must take off your boots, and carry them for a distance, or else the rats will follow you to your own house.

Thus, you know the whole secret of bewitching rats; and now I shall resume my remarks upon rat-catchers.

Again, I say, pay your rat-catcher honestly, and he will keep you free from vermin. What is six pounds a year to pay such a man? You may think it high; but it is no such thing. What is it? why, not two shillings and four-pence per week. Now, the great rat-destroyers of London inform me that with every economy it costs them at least ten shillings per week to keep a hundred rats; and I will give you my word for it that they do not feed them up like prize pigs, sheep, or oxen, to be exhibited at the Cattle Show. No; they give them just enough to keep them actively alive, and that is all. Now, at ten shillings per hundred per week, that is two shillings and sixpence per week for twenty-five rats. Do you think you have twenty-five rats about your farm? If so, why then, at that calculation, they cost you twopence per week more than the rat-catcher would, and all this is saying nothing about the young they would produce. But if we calculate them according to the wine-glass standard, and value the corn at only sixty-four shillings per quarter, these twenty-five rats would, in three weeks, eat one glass over one bushel and three half-pints. Then these twenty-five rats would cost you two shillings and eight-pence per week, which is fourpence per week more than the rat-catcher; but should you have five hundred, then, by employing a rat-catcher at the rate of two shillings and fourpence per week, you would save over twenty half-crowns weekly. Or, to give it in plain round terms, and averaging the corn destroyed to be worth sixty-four shillings per quarter, five hundred rats, to say nothing of their young, or of mice, would, in twelve months, by eating only a wine-glass each in twenty-four hours, cost £146. But, to estimate the rats at a thousand, the cost would be £292.

Now, do you think it worth your while to give a rat-catcher six pounds a year, not only to keep him honest, but to save you thus much money annually? If you have never given the matter a consideration before, it is well worth your most serious reflection now. I say, pay this man six pounds per year; and surely it shall be worth a farmer's while, when a London hotel-keeper pays the same merely to keep his house free from rats. However, by paying him that sum he would be able not only to live comfortably, but to supply himself well with dogs, ferrets, nets, traps, and poisons, and devote his entire time to about twenty farms. In that case he could give them each a monthly visit, and by so doing he would keep the vermin completely under. Not only that, the remuneration would have the effect of bringing a number of respectable scientific men into the market, and there would be an end to all losses by vermin.

Here let me tell you of one great blunder the majority of farmers make. When they employ a rat-catcher, they do not employ him as a protection against the ravages of vermin, but give him a job now and then, as a rat-butcher. They first go to the enormous expense of breeding and feeding a multitude of rats, and then, when they find their place completely overrun with a host of fine fat fellows, they give the rat-catcher a few shillings to come and kill as many as he can in a few hours. Why, of course, it is not this man's interest to kill all their rats; because if he does, there is an end to his work, and he may starve.

The way farmers should engage a rat-catcher is this. They should contract with him, as a scientific man, to keep them free from vermin; and by allowing him a fair salary (say thirty shillings per quarter), as a remuneration for his time, skill, and experience, he will find it to his interest to be vigilant and industrious. When the farmer finds himself comparatively free from rats, it will be a proof that the rat-catcher, like an honest man, is doing his duty, and therefore, instead of discharging him, as has too often been the case, let the farmer make him a handsome present, as a reward for his honour and industry; and take my word for it, he will kill every rat as it makes its appearance.

These men ought to be as actively employed in the summer, as they now are in the winter. For as the spring advances, the majority of rats leave the ricks and barns for the hedges, ditches, and water-side. There he ought to be frequently at work, with two or three well-trained dogs, to find out the holes; and into which holes he ought to put poisoned food. The consequence would be, that when the harvest was brought in, there would be scarcely any rats left to seek the barns and ricks for winter quarters.

Now, give these matters your most serious consideration, and I am satisfied that the more impartially you consider them, the more clearly will you see the removal of one of the greatest causes of misery and distress among agriculturists in general. As a proof, if each of you had by you now all that rats have deprived you of during the last ten years, what would be your present condition and circumstances? Probably you would all be rich men.

However, let me wind up this part of my subject by again adverting to the error among farmers in general, as to the time and manner of employing the rat-catcher. In most cases they never think of rats till they have become so numerous that they threaten destruction to everything around them. The ricks must be drilled full of holes, and the barns become a living mass, and yet it is not time to send for the rat-catcher, because he wants a few shillings more than the farmer feels disposed to give him. No, he must wait till the corn in the barn is thrashed out; and when it is thrashed out, instead of so many quarters of corn, as the farmer expected, he perhaps does not find one-half. The rats have eaten it. Then what is to be done to make up the quantity? Why, such a rick, which has been standing for years, and, according to the price of corn, is now worth so much, must be unthatched, to make up the desired amount for the market, because rent-day is at hand. Well, it is unthatched, and pray what do they find? That, in some cases, it is nearly half gone; and in others, there is scarcely a bushel of corn in the whole rick. This is no uncommon occurrence, and is too well known to nearly every farmer in the British dominions.

There is an old proverb, and a very true one, which says, "that if you want anything well done, there is nothing like doing it yourself." So thought ready-money'd Jack, who was his own rat-catcher. He told me he never did a day's work in his life; all he ever did, was to give orders, destroy the vermin, and attend the markets; and by these means he became a rich man.

CHAPTER VIII.

GOLDEN RULES FOR FARMERS.

Rule 1.—If a farmer would grow rich, let him be his own rat-catcher, and never let a rat rest till he has killed it. Though it shall cost him a day's pay, it will be pounds in his pocket.

Rule 2.—A live rat in the barn eats money—while a dead one in the dung-heap makes money; and remember this, that the death of a rat cuts off all the expenses of a worthless, hungry, multiplying family.

Rule 3.—If a farmer would sooner raise ricks than rats, let him down with his hedges, fill up his ditches, and drive the ploughshare through the rat preserves; and instead of rats, grow corn. But if he would turn rats into gold, let him fatten the soil with their carcasses, and he shall reap a golden harvest.

Here I would observe, that the average loss of land in hedges and ditches is estimated at one acre in twenty-five; that is, eight acres of loss on a farm of two hundred. But even in a closely-cultivated county, five acres have been gained, on a farm of two hundred and three acres, by rooting up the hedges.

Rule 4.—The old proverb says, "look after the pence, and the pounds will take care of themselves." But the fourth rule says, look after the rats, and the ricks will look after themselves.

Rule 5.—If you would go a joyous man to market, and return a rich one, build all your ricks on staddles, and give every man, woman, and boy warning, that if they put anything whatever against the stacks, or under them, so as to form a ladder for the vermin,—that instant they are discharged; if not, you will find one-half your grain turned into rats, which will feed and fatten on the remainder.

Rule 6.—By putting the above rules into practice, the careful farmer will turn his rats into guineas; while the sluggard, by laziness, will turn his guineas into rats.

As a proof of what I have stated in the last rule, it will be necessary to refer you back to two cases only. The first is of a farmer at Hanwell, who unthatched one of his ricks to decide a rat-match with dogs, for a thigh of mutton; and off which mutton he was to sup, as a return for his kindness. The dogs killed 170 each; and one of the owners, seeing nine rats make their escape into a piece of old wall, removed a stone, and drew them out by their tails, and let his dog kill them, thereby winning the match by nine rats. Now the rats that were killed from that one rick alone amounted to 349, which, at the wine-glass standard, destroyed in six months, or 182 days, no less than 127 bushels of corn; and this corn, at 64*s.* per quarter, would produce £50 16*s.*, which would purchase 152 legs of mutton, at ten pounds each, costing five shillings and fourpence per stone.

The truth is, that, with the exception of four days and Sundays, the corn these rats were destroying would have supplied him with a ten-pound leg of mutton every day in the week for six months; or, to make it more plain, they were costing him at the rate of two pounds per week every week they were there; and if this was the state of one rick, pray what must have been the state of his entire farm? Still, to what cause shall we attribute this folly and waste? The truth is, it matters little whether it arose from carelessness or laziness; there was the fact, that those rats were costing him at the rate of £100 a year; and though a rat-catcher was living close by, yet nothing induced this farmer to disturb them but a dog-match and a slice of mutton.

The next case is that of the brewer and farmer, who had his barn emptied for the purpose of trying a bull-dog with rats in a tub. In this case, none but the largest were selected for the tub, the rest being dashed against the wall, and so disposed of. In both these cases they knew they had plenty of rats, and knew also where to pitch on them, but would not disturb them till there was to be some sport. Now is it not a most extraordinary thing that neither of these men ever thought of destroying their rats for purposes of profit as well as sport. But had the depredators been discovered to be a drove of wolves, and had deprived them of a sheep weekly worth two pounds, what then would have been the consequence? Why the whole country would have been up in arms, and nothing but death and destruction to the whole of the wolf tribe would have echoed through the remotest corners of the island. But because the real robbers were rats, which from their numbers and daring were infinitely more destructive, little or no notice was taken of them.

CHAPTER IX.

HOW FARMERS AND OTHERS SHOULD EXTIRPATE THEIR VERMIN.

My plans for the extirpation of rats are very soon told. There wants but the will, backed with a little perseverance and determination, and the way will very soon be found both simple and easy. But bear in mind, that the first cost will be by far the cheapest; therefore set to work in downright earnest; and, with ferrets, dogs, sticks, traps, and guns, kill all you can, and poison the remainder!

In the first place, lay down a tarpauling to catch the loose grain, then down with your ricks, and rebuild them on staddles. But kill the enemy as they fly. As for those ricks that are already on staddles, just look them round, to see that there are no ploughs, harrows, hurdles, stumps, or arms of old trees, under or round about them, so as to form ladders for the vermin.

Yes, it is said, but rats will hang in the sheaf, as it is taken from one rick to the other; but, in answer, let me state, that I have seen a rick built nearly a yard from the ground, which was still swarming with rats. How was that, you may ask; why, on looking on one side, next the hedge, there was the arm of an old tree, half broken off, resting against the rick, and forming a ladder. I drew the farmer's attention to it, and bade him have it instantly sawn off. The consequence was, that during the night the rats were compelled to jump down for water, or die where they were; but when down, they could not get back again; and so the rick was cleared of them, and this will be the fate of those rats that cling in the sheaves.

I've seen another instance of a rick built on staddles being drilled with holes. But in that case the farmer had built a haystack against it, and the rats had worked their way through the stack into the rick. Indeed almost endless cases might be adduced where ricks, though standing on staddles, have still been infested with rats, and which fact has often led farmers to make the observation, that it's no use building your ricks on staddles, because, whether you build them on staddles or on the ground, it's all the same; for when the rats want them they'll have them, in spite of all you can do. This is perfectly wrong, and not consistent with common reason. I will undertake to say, that in no case whatever did rats ever get into a rick that was properly built on staddles of a sufficient height, unless there was a ladder of some kind or other for them to get up and down by. The thing is a physical impossibility.

A rat may sometimes be seen lapping the rain on the top of a rick; and I have also seen the thatch, after a shower in hot weather, half covered with mice catching the drops of water as they run down the straws. Mice will also quench their thirst with the night dews on the thatch; but rats, like men, require something more than dew-drops to quench their thirst. The fact is, rats by nature are very thirsty animals; and, in addition to that, the dry nature of their food, in barns and ricks, renders them doubly so; the consequence is, that if they do not get a plentiful supply of water they will die; but I believe, that if a rick had a tent covering fixed over it, so as to keep off the rains and dews, it would have the effect of destroying all the mice. Nevertheless there are cheaper means of getting rid of them, of which I shall speak presently under the head of "Poisons." So let us proceed with the first step for getting rid of the rats.

My rule is, down with the ricks, and rebuild them on staddles. And now, let me ask, what would be the expense of such a job? Of course a great deal depends upon the size. However I have made myself pretty well acquainted with the cost of such an alteration. A good-sized rick takes eight men two days to take down and rebuild; and averaging their wages according to what they are mostly paid, namely, ten shillings per week, it would cost £1 6*s.* 8*d.*; and allowing seven shillings for thatching, then the entire expense would be £1 13*s.* 8*d.*

Now suppose you destroy but thirty rats—never mind—you will be immense gainers by the outlay; for if only one-half be females, pray, in six months, what would be the number of their progeny? Let us see: fifteen females, at eight young ones in a litter, would produce 120 the first littering, and in the second littering there would be 120 more, making together 240 young. But this is not all; for the first brood would now be coming in six months old; and, supposing them to be half females, they would have among them no less than 480 young ones. The consequence is, that, with the first thirty, you would, in six months from the first littering, have no less a number than 750 living rats; and which living rats, at the wine-glass standard, would consume one bushel and a half of corn daily, or ten bushels and a half weekly; and valuing the corn at only sixty-four shillings per quarter, these rats would cost twelve shillings per day, or £4 4*s.* per week, every week they were with you.

Now is it not worth your while to rebuild your ricks on staddles, at so small an outlay as £1 13*s.* 8*d.*, to save four guineas weekly? Or will you let them remain to ruin your ricks, and after that, to go in swarms to your neighbours' ricks, and ruin them also?

Now that I have succeeded in convincing you to some

extent of the heavy losses you sustain by vermin, I must tell you what becomes comparatively easy and amusing. If the farmers of England would form themselves into neighbourly bodies, for the purpose of carrying out the war of extermination, three months would scarcely pass over, before rats, ay and mice too, would become nearly as scarce as kingfishers. This ought to be done, because the rat is the common enemy of all; and though he may be destroying your neighbour's property this week, perhaps next week may not pass over before he is destroying yours; and it matters little whether you kill him on your own or your neighbour's grounds; you both become free of him.

If this plan were strictly followed out in every district in England, then there would be scarcely a rat left to travel from farm to farm, parish to parish, or from county to county. But whether farmers in general combine together or not, it will not prevent each farmer looking after his own individual interest, and acting for himself.

Now suppose you have succeeded in building a rick on staddles, and cleared away everything both from underneath and round about which could in any way form a ladder for the vermin. But farmers should continually be on their guard, because there is no knowing what ill-natured people will do, particularly disappointed rat-catchers, who are often at their tricks. Should you at any time find anything, remove it directly; and though they may have put some rats into the rick, never mind, your removing the ladder will directly get rid of them. But before proceeding to the barn or granary, I will tell you of two newly-invented traps, which, in my humble opinion, will lead to the capture of more rats than all the other inventions put together, and these are Uncle James's infallible rat-traps. These traps, if properly applied, are warranted to render every rat in the premises a living prisoner in forty-eight hours, but mostly the first night, if the directions be faithfully carried out. They entirely supersede every other invention, as also all kinds of poisons, and thereby do away with the possibility of accidents arising out of the use of such deadly drugs, particularly in dwelling-houses. They need no baiting whatever, and a child may set them with perfect safety! but when once set, they never after require touching, unless removed to another place; and whether there be but one rat or a thousand, it is all the same; they make prisoners of all, without hurting any. But, independently of their humanity, they also do away with the possibility of that most detestable of all nuisances—rats dying and rotting beneath the flooring and behind the wainscoating of our premises. They are equally effectual, whether in barns, granaries, storehouses, warehouses, dwelling-houses, or on board ship, or in cellars, sinks, drains, workshops, or garrets. Night or day they are ever ready, and act without the possibility of a failure. They need only to be seen, to prove their infallibility, and the most inexperienced person may at once capture rats with as much skill as the first ratcatcher of the age. They are equally as valuable to the merchant and the mariner as they are to the farmer and the tradesman; and they are not only an invaluable panacea for all in England who may be afflicted with rats, but a boon to the human family, since they are applicable to every species of building in every clime beneath the canopy of heaven. Their simplicity and infallibility are their principal recommendations and virtues, and one trial only will prove to a demonstration the truth of what I have stated. All they require is a clear field for action.

Nevertheless, for the satisfaction of those who may still have their doubts as to the quality of these traps, and feel a prejudice in favour of other plans, I will, for their benefit, give the best means for baiting traps, as also the most approved poisons, with the best method of mixing and using them; and then they can take their choice. But, for our present purpose, we will suppose you have gone to the expense of procuring a set of Uncle James's infallible rat-traps. Having secured the rick, direct your energies to the barn; and, if it be much infested, first stop all the outside holes with stones, brickbats, or coarse gravel; and if there be any crevices in the sides through which the rats may escape, stop them up closely with hay thrust hard in with the point of a hedge-stake; that done, empty out the corn, and sweep the floor, in order to find out all their holes; and as you find them, fasten a trap over each, and see that it works clear and easy; that done, place the box-traps in the corners. When all the holes are secure, and everything in readiness, put in the ferrets, and then stand still till the rats bolt out; but do not attempt to kill them, unless they are crawling up the sides; in that case, quietly cut them down. An ash stick, about the thickness of your little finger, is the best weapon. As for the rest, they will run into the box-traps. There you will have them all alive, and can dispose of them as you please. A youth, in this way, may clear a barn in an hour or two. But when all is over, and the ferrets safe, just put a few phosphoric balls down the holes, in case there may be a straggler or two; and that will send them to the nearest water, where they will drink till they die, and there will be an end of them.

The barn being clear, take the traps and ferrets to the granary, and treat it the same. But in either case, should there be but a few rats, it will not be worth the trouble of clearing out the corn. In that case feed them for a few nights with scraps of cheese and bread; but be sure to put down some low dishes, and fill them to the brim with clean water for them to drink. Do everything gently and kindly. By doing this for four or five days, so that they may know where to run to for water, they will eat the food freely; then put down some phosphoric balls, or phosphoric paste, spread on bread; this renders them so thirsty, that they instantly run to the water, where the more they drink the worse they are, and in most cases die beside it. This is worth all the other poisons put together for destroying them in buildings, because it is less dangerous, and more effectual; since you can throw out their carcasses, and thereby do away with their stinking about the premises. But whatever you do, let them have plenty of water; if not, they may crawl back to their holes, and there die. These balls are excellent things for the hen-house; but be sure you put them far enough down the holes, so that the fowls cannot pick them out, and that will preserve your eggs and poultry.

As for the cow-house and stables, treat them the same as the hen-house.

The next consideration is, how to clear dwelling-houses, drains, &c.—In any apartment, where, by shutting the doors, you can cut off every means of escape for rats, except down their holes, the ferrets are not actually necessary. The traps alone are sufficient, because rats cannot or will not go forty-eight hours without food or water; consequently the instant they poke their heads out of their holes, in the search of either, they are prisoners; and in the morning nothing remains to be done but to destroy them, which may be done by a dog, or with a stick; or if you have a box-trap in the corner of the room, the instant you enter they are most likely to run into it to escape; and then you can do with them as you please. But some one will lay down poisoned food and water, in order to save the trouble of killing them. In that case you need only pick them up and throw them out, or bury their carcasses in the dung-heap. On the other hand, should there be any cracks in the flooring, or under the skirting where the flooring has given way, and in and out of which the rats can run at pleasure, you must stop the places with wood, or any other material, so that they cannot get in and out of any place but their holes. If not, you render the traps useless; and in that case all I can recommend is, to catch them one by one, with any of the ordinary traps; or else bait the place with poisoned food and plenty of water, in large plates or low dishes, and keep the place very quiet, because, if they are much disturbed, they may have just strength enough to crawl away, but not to get back again; consequently they would die and rot in their holes; and how often are people compelled to have the flooring taken up, or the wainscoating removed, to get at their putrid carcasses? This again proves that the first trouble or expense is by far the least.

In the event of your using Uncle James's traps, never fail to go the first thing in the morning, and get rid of the prisoners. If not, and you neglect them, they will have time, and may gnaw their way out, and then you will have to put a trap over the new hole. But I must tell you, that I never knew a rat to gnaw a hole through the flooring or skirting in one night. It is mostly a work of time.

Again, should you have any doubt as to whether there are any rats in a particular place, it can soon be ascertained by sifting a little sand around the holes. You must lay it on very thin, and in the morning you will see their marks, if there be any; but if there are none, the sand will remain the same as you left it.

While the work of destruction is proceeding on your premises through the night, you can be employing yourselves in the grounds through the day, by hunting out all the rat-holes in the hedge-rows, ditches, and the water-side, and indeed, anywhere and everywhere where you are likely to find any. Then, as you find them, put into each some poisoned food. The best implements to use are a decent-sized milk-can, with a lid to it; let it hang at your side by a strap passing over the shoulder. It should be kept for the purpose, and if it were made to fasten with a padlock it would be all the better, as that would prevent accidents. Then to convey the poison from the can to the holes, you should have a large, deep table-spoon, fastened to the end of a stick,—say a broomstick; but be sure to put the food far enough down the holes, and then no harm can come to any of the farm-stock. You must not fill the spoon too full, for fear of spilling. You should hold it over the can, and with a knife or piece of wood make it strike measure. Two of these will be enough for one hole, and as rats run from hole to hole during the night, those from the neighbouring holes you have missed will be sure to find it, and so you will get rid of them. This plan is both simple and easy, at the same time most effectual. By going the rounds a few times during the summer, and each time poisoning the holes, the result will be, that by the autumn there will be scarcely a rat left to come into the farmstead. You will thereby not only reap the full benefit of your abilities and industry, but the country also will enjoy the full blessings of a bountiful harvest, both of which are, of all earthly blessings, the greatest next to health.

There is another point to which I wish to draw your particular attention; I mean the old drains about your premises, which form most safe retreats for rats to live and breed in. I would recommend you to do away with all those that are useless. But those that are absolutely necessary you should put into good repair, and have them secured at each end with sliding gratings dropped into grooves; so that at any time you can lift them up, either to sluice out the drain, or remove any obstruction, and then replace them. This will save a great amount of trouble and expense, besides most effectually keeping out the rats. But of all things, for drains, there is nothing, in my opinion, equal to pottery drain-pipe, not only for cheapness and durability, but because they are proof against the teeth and claws of rats; or, more properly speaking, they are perfectly rat-proof.

CHAPTER X.

THE POLICE OF NATURE.

Having laid down the various plans of operation for the entire destruction of rats in every portion of the farm and out-buildings, I shall now devote a few pages to what are generally looked upon as the natural extirpators of the rat; and hence called the Police of Nature.

The common enemies of the rat are, for the most part, the marten cat, polecat, stoat, weasel, cat, dog, owl, and man. We shall give each a fair consideration, and see how far they tend to destroy rats; and perhaps we shall find many receiving far more credit for their exertions than they merit.

In the first place, we shall take the ferret, which is a native of Africa; but here he is, comparatively speaking, a domestic animal, being bred by gamekeepers and rat-catchers for rat and rabbit-hunting.

Ferrets were first taken from Africa to Spain, for hunting out the rabbits, which had become so numerous that they were producing a famine, and threatened destruction to the country. From thence they spread throughout Europe.

When ferrets are wanted for hunting, their masters keep them without food, because they will hunt but little, if at all, when they are full. Consequently ferrets hunt not for the purpose of destroying their victims, but for the sake of a meal; and when they are put into the holes, if they should slip their muzzles, and catch a rat or rabbit, or come across a nest of young ones, they will destroy them; and after having gorged as much as they please, will often lie down by the remainder, and go fast asleep! As for rats or rabbits, they may do as they please; the ferret will sleep on, being naturally a drowsy animal; consequently the rats destroyed by ferrets themselves are very few, considering they do not run wild in this country, whatever they may do in Africa.

Here I must remind you, that those ferrets that have been brought up to rabbiting are seldom of any use for ratting. The reason is plain. In rabbit-hunting they are in the habit of having everything their own way, and never receive a bite for their trouble. But when put to ratting, and they get punished, they often bustle out much quicker than they went in; and in many cases will never enter another rat-hole as long as they live. I have been let in by gamekeepers and their ferrets in this way. Some of them you could not get into a hole, unless you crammed them in; and then it was like cramming a cat into a boot; as for hunting, that was out of the question. Then others, again, were cowed the first time they were punished, and were good for nothing after. The consequence was, I resolved never after to buy either dog or ferret, till I saw them fairly tried; and if you take my advice you will do the same. But I must tell you that I never yet saw an old ferret, however good, that would face a full-grown rat that had backed into a small hole. He will work his head to and fro as if giddy; and the instant the rat turns, he will attack him; but never while the rat is facing him with open mouth. I have had resolute young ferrets attempt it the first time, but never the second.

The polecat seldom or ever sees a barn rat, because his hunting-grounds are mostly in rabbit-warrens or game-preserves. This is well known to gamekeepers, who find him the worst poacher they have to contend with. Besides, he is too large to get into the majority of rat-holes. In addition to that, he is naturally wild and solitary in his disposition, and gets as far as possible from the abodes of man. Nevertheless I have heard of his being found about the outhouses, and even on the top of the farmhouse itself. But when he does visit the farm, he commits great depredations, if not speedily destroyed. I have heard of cases where he has destroyed every pigeon in the dovecot in one night, and in others, where he has committed sad destruction in the hen-house. But mostly his food is game of any description he can get at, and among which he is very destructive, frequently taking life merely for the blood or brains, or to gratify a murderous propensity. Gamekeepers are ever on the alert, and as soon as they discover him to be in the neighbourhood, they never let him rest till they have killed him; the result is, that there are but few polecats to be met with; consequently they can do but little towards destroying rats, even were they so disposed.

The stoat is somewhat like the polecat in disposition, and very destructive among partridges, young pheasants, leverets, rabbits, and chickens. Sometimes, when hard pressed for food, he will hunt the water-rat; but if the rat gets the start, he disappears in the water, where the stoat feels but little disposition to follow him. Game is his general food, and gamekeepers never let him rest till they have killed him; consequently his ravages among rats must be very limited.

The home of the weasel is anywhere, but principally in hedge-bottoms and old walls. Sometimes he will come about the farm-buildings and hen-roost, and is very destructive among chickens, for which farmers and their men mostly kill him wherever they can; but he will at any time give the preference to a chicken or duckling, to fighting with a rat for a meal. He is generally too small to contend with a rat; and I have my doubts as to whether a weasel could beat a full-grown rat in a fair open contest; for I know of no animal of the rat's size, except the squirrel, which can inflict so deep a wound.

A celebrated naturalist tells us that while he was watching some birds by the copse side, a hare ran out not far from him, and, after pausing for a minute, returned through another opening. The hare had scarcely entered the copse, when out came a weasel in pursuit, tracking the hare by scent, and disappeared again in the copse. Presently out dashed the hare a second time; then threw itself down in the grass, and screamed till the weasel came up and seized it in the nape of the neck. The struggle was but of short duration, for before our authority could get to the place, the weasel had run away, and the hare was dead. This almost sudden death of the hare not only surprised him, but excited his curiosity to examine the wound, and upon close inspection he discovered only two little punctures, like small pin-holes;

but beneath, there was a black spot of congealed or coagulated blood, about the size of a shilling, occasioned by the weasel's sucking. Hence it is tolerably clear that hares and rabbits, when attacked by weasels, do not die of the wounds themselves, or from a fracture of the skull, as is generally supposed, but from a species of apoplexy, or rather a stagnation of the blood on the brain, occasioned by the congealed blood stopping the circulation.

On another occasion, two skilful bird-catchers were employed by a bird-fancier to go into the country for a week, and catch him all the bullfinches and goldfinches they could, for his shop. From Monday to Friday they had excellent sport, and had caught a number of the finest birds; and, as might be expected, were in high spirits at their success. The hens were put into one store cage, and the cocks, carefully selected, into another, where they received every possible care and attention, to preserve their health and plumage. On the Saturday they were on the ground by daybreak, resolved to catch all they could up to midday, and then start by the train, so as to be in London by the evening. The store cages, as usual, were placed in the ditch, with canvas over them, to prevent the birds being startled, and dashing about the cage. The morning turned out most unfavourable and wet; therefore, they packed up their nets, and resolved to start for London by the first train. However, upon going to the cages, and uncovering them, the hens were sprightly enough; but the cocks were every one of them dead! The grief and astonishment of the men are more easily conceived than described. But what could have killed the birds? It was soon explained; for upon looking into the cage, there lay a little weasel in one corner among the dead. It just popped up its head, as much as to ask what business had they to disturb it in its slumbers. The men jumped about like madmen after stones, to kill it; but before they could succeed, the little murderer worked his way through one of the holes where the birds put out their heads to drink, popped through the hedge, and got clean away. Thus ended this tragedy, though with but little sympathy for the men; for what right had they to imprison and enslave these pure songsters of the woods for mere profit and pastime?

My motive for mentioning these circumstances is merely to show the habits and disposition of the weasel; at the same time to prove with what ease and security he can obtain food without contending with rats, and you may rest assured that he has no more liking for punishment than we have. At the same time it shows what a destructive animal it is, when a little creature like that, which could not have been much bigger round than a mouse, was not contented till it had taken the life of every bird in the cage. The same destructive disposition will be manifested, when it can get among a brood of chickens, ducklings, partridges, or pheasants, unless disturbed. They hunt either day or night, and rest only when tired. Their general food is hares, rabbits, leverets, small birds, and mice; and they have no objection to a turn in the hen-roost, if opportunity offers, or to ransack a pheasant or partridge's nest that may by chance come in their way; but they will never attack a full-grown rat, unless pressed by extreme hunger; and then they have been known to attack not only children but men also.

The truth is, that polecats, stoats, and weasels do very little towards destroying rats, as they seldom frequent the same localities. Besides, the gamekeeper's special business is to preserve the game; consequently he will kill anything and everything that tends to destroy it. He is constantly on the watch, and what with dogs, traps and gun, he will never let these poachers rest till he has put an end to them; and besides, there have been for ages rewards given by the parish officers for their heads, dead or alive. The result is, there are, comparatively speaking, very few polecats, stoats, or weasels to be met with.

The marten, better known as the marten-cat, is a very beautiful animal, whose skin is much sought after by furriers, being considered almost equal to sable. Consequently the hunters have ever been on the alert for it. This, with the gradual reduction of forest lands to a state of cultivation, has, in most counties of England, rendered these animals almost extinct.

Like the squirrel, the marten lives mostly in trees; but, instead of nuts, it devours all kinds of birds that come within its reach, and seldom comes to ground, except to quench its thirst, or plunder the hen-roost of some lonely farm. But to see a marten is a great rarity. Consequently their extreme scarcity must of necessity render their services very trifling towards thinning rats, even were they so disposed. The truth is, that so far as martens, polecats, stoats, and weasels are concerned in the destruction of rats, we might very easily dispense with their services. But when we take into consideration their ravages among all kinds of poultry, the sooner you kill them, and send their skins to the furriers, to be made into muffs and tippets for your wives and daughters, the better; since they will render far more service in furnishing winter covering than in plundering the farm.

Nevertheless, from the scarcity of these animals, I am satisfied that the great majority of outrages committed in the poultry-yard, for which foxes and the weasel tribe are condemned, are the work of rats; not that weasels, &c., are the more to be pitied, or the more to be trusted, since they want but the opportunity to substantiate all that can be laid to their charge.

So much for the Police of Nature. And now for Miss Puss. I speak of Miss Puss because Master Thomas is so drowsy a beast, that he sometimes will not stir even for a piece of rumpsteak, though it shall be nicely cooked and buttered for him. Talk to him of killing rats! Let those kill rats that like it. For his part he has enough to do to look after the fair sex, without being bothered with rats. But the habits of the cat are so well known, that they require but little comment from me, except that they often receive far more credit for rat-killing than they merit. Through the day they are in a state of laziness about the fireside, and in the evening they are looking after some larder, birdcage, or aviary, for something to eat; and after that, instead of hunting for rats in the cellars, they, to the annoyance of the whole neighbourhood, are attending the midnight concerts on the house-tops, and leaving the premises for the rats to look after in their absence. But where there are many rats, cats frequently become so used to them, that they will not only not touch them, but have frequently been known to feed with them. Still, some cats are determined creatures, and will watch for days together at a rat-hole. Sometimes they catch the rat, and sometimes not; but look at the waste of time! On the other hand, rats have keen noses, and can smell the cat; so they shift their quarters to some other place of greater safety. The result is, that rats destroyed by the cat tribe are very few, because they cannot get into the holes after them; but were they small enough, I doubt their courage; for I have seen a large cat dead beat by a rat in a fair open contest, when the latter has been allowed to walk deliberately to his hole.

But this I can tell you, that some cats are most destructive creatures among pigeons, rabbits, and chickens. These are facts in which I have been a sufferer. My first lot of pigeons I lost by rats! After that, when living in the neighbourhood of Oxford, I had another flight of fancy pigeons, comprising twenty birds in number. All went on very well for a time. Presently I missed one; the following morning I missed another; and so on for about a fortnight, when I found I had lost fourteen out of the twenty! Of course it created a stir in the place, as to what had become of them; and the neighbours began to suspect each other. However, a man came and told me that he saw a cat run up the road with a blue pigeon in its mouth not an hour before. This was the first thing in the morning before the people were stirring. We traced the feathers for some distance. When evening came, I loaded my gun, and requested them to wake me up at daylight. It was summer time; so I left the window a little open to be ready. First thing in the morning they came to my bedside, and told me the cat was there. Out I crept to the window, and there I saw the gentleman,—a big tabby Tom cat, crouched between the leaves of a large sunflower. The pigeons were picking about the ground. I gently put the muzzle of the gun out, and waited for him. Presently one of the birds came near; his tail began to move; and just as he was in the act of springing, I fired, and fortunately missed the pigeon, but hit his lordship in the hind-quarters, and fatally stopped his career. Still, before I could get out, he had scrambled away. However, four or five days after, a little old woman came, inquiring if any one had seen her Tommy? He was such a faithful creature, and always came home at breakfast-time to have some milk. Indeed that was all she ever gave him, because he got his own living!

And this is the way that thousands of people keep cats at others' expense. But, unfortunately, cats are only held as vermin; consequently there is no law to make their owners responsible for their depredations.

On another occasion, in the place where I kept my pigeons, some one had the kindness over night to leave the door open; and upon my going in, in the morning, a great carroty Tom cat waddled out, and made its escape! Now whether he had had any one to supper with him, I know not; but he left me only the legs, wings, and feathers of five choice tumbler pigeons.

Not long after that, my son had a handsome young rabbit, about three months old; it was considered a choice specimen of the fancy lop-eared breed, and was very much prized. However, one morning a hungry half-starved looking cat managed to get at it, and bent the wires of the hutch in dragging it through. I found her in act of eating it, though the rabbit was nearly as big as herself.

As to chickens, I, on one occasion, at some trouble and expense, managed to get five eggs of a very fine breed of Spanish fowls, and succeeded in hatching five chickens both sound and healthy. When they were a fortnight old, I loosened one of the bars of the coop, to let them have a run in the garden without the hen. I was much delighted with their appearance; they seemed sprightly and vigorous. After a time I went in to dinner, and left them to enjoy their run among the flowers. However, upon my return, I missed two; and while wondering what had become of them, down jumped a great Tom cat, and seized a third. Over the wall he bounced with the chicken in his mouth, and away he went before I could get anything to send after him.

Thus it is quite clear that large cats are just as destructive among pigeons, rabbits, and poultry, as any other carnivorous animal.

The most useful cats for farmers, are little, shy, agile, and mostly glossy black, with a small flat head, large eyes, and longish ears, are excellent mousers, and will often fight a rat with great determination. These small cats should be taken great care of, and the breed encouraged, as they are most valuable in houses, barns, and granaries. As for the larger cats, the best thing you can do, is to get rid of them; for they are too lazy for mousing, and too cunning for ratting, and will at any time sooner dine, sup, or breakfast off a pigeon, chicken, or duckling, than battle for a meal.

We now come to the farmers' most sincere friend, the BARN OWL, which, for beauty and softness of plumage, and delicacy of colour, is scarcely to be surpassed by any bird in the entire range of ornithology, and which, at the same time, has been the most barbarously treated bird in the whole creation. Country fellows have taken the poor birds' eggs, and thrown them about, seeing how far they could sling them before they would break. Then they have taken the half-fledged young ones, and after subjecting them to every cruelty, by knocking and kicking them about, they have twisted their necks, and thrown them into a ditch. Then, again, savages, calling themselves men, have taken one of the old ones, and having lashed it to a duck's back, sent it adrift in the pond, and there let it be till it was drowned by the duck's continued divings to get away; and the struggles and the piteous cries of the poor purblind bird for life were by these ignorant barbarians considered subjects for merriment and laughter; and lastly, the poor bird's companion was spread out against a barn-door, and nailed through the head and wings as a warning to others. Thus have they disposed of this innocent, and, at the same time, most valuable family.

To justify such barbarities, pray what are the charges brought by these rude savages against these unoffending birds? That they suck the pigeons' eggs, and kill the young pigeons and chickens; and that they devour the smaller kind of birds, and plunder their nests; each of which charges I believe to be entirely false. In substantiation of this belief I will quote the testimony of that most interesting traveller and naturalist, Charles Waterton, Esq., whose evidence in such matters I hold to be a host in itself, because he is not a theoretical, but a sound practical naturalist, who spares neither time, trouble, nor expense to arrive at pure natural facts. He tells us that when farmers complain that the barn owl destroys the pigeons' eggs, they lay the saddle on the wrong horse; they ought to put it to the rats; for that he formerly could get scarcely a young pigeon for the table; but after he found out the cause, and had all the rats in the dovecot destroyed, then every year he had young pigeons in abundance, although some barn owls regularly roosted in the dovecot, and others were and still are encouraged all around it. Yet, so far from the pigeons feeling any alarm or distrust, they seem perfectly reconciled to them. But should a hawk make its appearance, then in an instant the whole colony is up. He tells us also, that he has been repaid a hundredfold for protecting and encouraging these beautiful birds by the enormous quantities of mice they nightly destroy. They will also devour beetles, and other large insects that come in their way; but mice are their principal food, and the food they hunt after.

As to the rats destroying pigeons' eggs, and the young pigeons, I can bear testimony; for a colony of them ate not only my eggs and young ones, but the old ones too. Yes, in one night they devoured my whole flight; leaving nothing but shells and feathers behind; and I must tell you also, that among them there was a pair of fancy tumbler pigeons' eggs, for which I would not have taken a five-pound note. You will read the whole account under the title of "Cruel destruction among rabbits, pigeons," &c.

Then to prove the barn owl's inability to hunt in hedges and the close branches of trees after small birds or their nests, just put one into the middle of a bush, and you will soon see that it is as helpless as if it were in the water. Nature has not framed it for such work. The eyes are too large, and the feathers upon the body, head, and face, are too soft, and not rightly made or placed, to enable it to glide through close quarters with the snake-like ease of other birds, whose every feather is pointing or bearing towards the tail. Indeed, if you examine the feathers on the owl's head and face, to say nothing of their extreme softness, you will find them so placed, that if he attempted to pass through a hedge, he would receive injuries similar to what another bird would by being dragged through the hedge tail first.

The plumage of the owl is very thick, for the purpose of defending it from the cold night air, and also very soft, to enable it to fly with such down-like silence, that even the quick ear of the mouse cannot hear it coming. This extreme silence, while flying, is caused by the formation and texture of the feathers, which, unlike other birds, are very loose, and fringed at the ends, and completely prevent that rushing sound which accompanies the flight of most other birds. Here then do we find it framed by nature for gliding over the grounds and ricks at twilight in quest of mice to furnish food for itself and young, and not for bird-hunting in the bush. They may sit on the big branches of a tree, near to the trunk, or get in a large hole in the trunk itself; but small birds seldom build or roost on or in either. As for chickens, they, like the majority of birds, are gone to roost long before the owl comes out. But should he by accident take flight in the open day, he gets a most unmerciful mobbing for his trouble by all the birds in the district; and what makes the matter worse for him, he cannot guard against his assailants, because he is too blind to see them! Consequently his habit is to remain quietly in his hiding-place till evening comes, and the coast is clear.

Nevertheless I have seen the owl eat small birds, but that was in the bird-shops of London, where they had them thrust into small cages, and fed with dead linnets. Still we must bear in mind that they were not in a state of nature, but closely confined, and seldom got a meal except when a linnet died; and that they often went a day or two without food; consequently it became one of two things, either eat the linnet, or die of starvation. But if a boy chanced to bring them a dead mouse, that was a delicious morsel.

Mr. Waterton tells us that he has often been glad to shoot a monkey for dinner; and, to use his own words, he says that, "a grilled monkey is not to be sneezed at by a starving man." Then, again, the *Times* newspaper tells us that a British officer, in the late war with Russia, only two days before the surrender of Kars, gave six-and-twenty shillings for a rat to eat! Indeed I do not believe that any one would turn up his nose at a roasted owl, under similar circumstances; though it is certain that we would not, on any conditions, eat them at the Lord Mayor's dinner. Neither does Mr. Waterton eat monkeys at his own house, or the British officer eat rats at the royal table. Nor would owls eat birds if at liberty. But different people have different notions with regard to starvation. It is one of those things that is far less painful to talk about than experience. For instance, the following conversation is related of one of the princesses in George III.'s time, when there was a great talk of famine and starvation. "Do you believe, governess dear," said the princess, "that it is possible any one could starve to death?" "O yes," replied the governess. "Well, I really can't believe it," said the princess; "for, do you know, governess, that I could sooner eat bread an' cheese than die of starvation." This, it appears, was the princess's impression upon the subject. But certain it is, necessity knows no law; and had misfortune placed her highness with Mr. Waterton in the wilds of South America, or with the British officer in famished Kars, it is ten thousand to one but she would have been as anxious for her share of the monkey or rat as any one else.

But let us return to the barn owl, in order to refute the unjust charges brought against the tribe. You must understand, they now and then disgorge, or vomit up, an oval ball or pellet, which is composed of the fur and bones of mice, that remain in the crop or stomach after the flesh and dissolvible particles have been disposed of in the ordinary process of digestion; but the fur and bones being difficult of digestion, nature has provided their crop or stomach with the means of working and compressing these substances into a ball, and then ejecting it. Of these pellets the birds make their nests, wherein they lay and hatch their eggs. Mr. Waterton tells us that he has dissolved and examined these pellets, and could never find either feathers or birds' bones in any of them. Now, as feathers are equally as hard of digestion as the fur of mice, and birds' bones no easier of digestion than the bones of mice, the only just conclusions we can arrive at are, that the barn owl, in a state of nature, never eats birds of any kind. Consequently the charges against it are unfounded; and the persecutions it has suffered have been most barbarous and unjust. But, on the other hand, Mr. Waterton says you can find the skeletons of several mice in each pellet; and that in one of his owls' roosts he gathered up at one time a bushel measure nearly full of these pellets, that had been disgorged within the last few months.

Here Mr. Waterton and nature establish two most valuable facts: first, that mice are the natural food of the barn owl, among which it is very destructive; and, second, that in a state of nature it never destroys poultry, or birds of any kind; for if it did, the remnants of feathers and birds' bones must of necessity be found in the pellets. This is a matter which any farmer, having owls about his premises, can test for himself, by dissolving the pellets in warm water. But until he has tested them, and can prove to the contrary, let him esteem the barn owl as his best friend and protector against the whole of the mouse and shrew tribes.

Again, when we consider the fact that Mr. Waterton gathered up in one roost nearly a bushel of pellets, that had been deposited there within a few months, and that each pellet contained the skeletons of several mice, pray what must have been the amount of mice destroyed in that time? Now let us suppose the juices and action of the owl's stomach to soften and compress the bones and fur of a mouse into barely a quarter of the animal's original size, what fact would it then establish? Why, that these few owls, in those few months, had destroyed over four bushels of mice.

The author of "British Ornithology" says that at twilight the barn owl may be seen issuing from its retreat to the adjoining meadows and hedge-banks in search of food, hunting with the greatest regularity, and pouncing upon its prey with deadly aim. This it swallows whole, without any attempt to tear it in pieces. He also observes that the owl is of most essential service in checking the increase of the various species of mice and shrews upon which it subsists. He says it is easily domesticated, and will become very tame, if taken when young, and kindly used. Montagu reared together a barn owl, a sparrowhawk, and a ringdove, who lived in great harmony for six months. They were then set at liberty; but the owl was the only one of the three that returned.

Now let me quote another able authority in behalf of the barn owl's valuable services to farmers. The author of the "Natural History of Selborne" tells us that he once watched a pair of barn owls that had a nest of young ones in the roof of the church. He says they went out at sunset in search of mice, which begin to run about at that time; and he saw the owls quarter the ground with the greatest regularity, like well-trained pointer dogs; and from their frequent returns to the church with mice, he had the curiosity to time them; and he found that they brought a fresh mouse to the nest about every five minutes.

He likewise noticed their peculiar mode of devouring a mouse, when they chose to have one for themselves. He says they first give it a sharp bite or two, as a terrier would a rat; then jerk it up, and catch it head first in their mouth; a second jerk sends it half down their throat, with only the tail sticking out at one side, like an old woman smoking her pipe; and after having amused themselves this way for a time, they give it a third jerk, when it disappears altogether. Then the bird seems perfectly happy, till it starts off for another mouse.

Still it must be observed that peculiar birds, in peculiar places, and under peculiar circumstances, like peculiar individuals in the human family, may have their peculiarities and propensities, which often lead them out of the ordinary course, and prompt them to do things contrary to the order of nature. But it is anything but justice to condemn a whole race for the freaks of an individual. For instance, Mr. Waterton relates a case of one of his owls that had a propensity for fishing. He one day saw some fish scales in one of the owl's nests. Curiosity prompted him to watch them in the evening; when he saw one of them stoop, while passing over the lake, and carry away a fish. He directly went to the nest, and found a very nice perch, which he took away with him.

The author of the "History of British Birds" tells us, that the remnants of small birds have at times been found in the pellets. All this no doubt is true; but I believe that the finding of the remnants of small birds in the barn owls' rejected pellets, is a thing almost as rare as the finding of fish scales in their nests, which fact Mr. Waterton considered a matter of such great curiosity, that he published it to the world. But all this does not prove that the owl destroys eggs, pigeons, or poultry.

Now let us suppose that one owl in a hundred may now and then, either wilfully or accidentally, pick up a lark or any other small bird that has dashed out of the bush in the night, from having been disturbed by some rat or weasel; or that they pick up the remnants of a little chicken which has been destroyed by one of these animals. Then let us place them in one scale along with all the fish Mr. Waterton's bird may have caught. After that, let us put into the other scale all the mice the ninety-nine have destroyed! What then would be the verdict which reason and justice would compel us to return? Why, that in addition to all the fish and small birds they would like to indulge in, they ought to have gold collars with brilliant appendages hung around their necks, for the immense services they have rendered to farmers. But they need no such gaudy trappings; all they require is mere protection through the day from their common enemy and persecutor—man! and then they will render him infinitely greater services than they have hitherto done. Thus in the night, if your wife or family should hear them hoot, instead of being alarmed, let them feel satisfied that the owl is that good and faithful servant who is up and protecting his master's property while the master is sleeping. We know that the apparent size and grotesque appearance of its head, and its being a bird of the night, and foolishly associated with all kinds of enchantments, has rendered the owl odious in the eyes of many! But then we should bear in mind that this antipathy is merely the result of ignorance and superstition. God has made him a bird of the night, not for the purpose of aiding incantations, but to watch the mouse tribe, so as to prevent them increasing till they eat man out of house and home. If those who entertain a foolish prejudice against the owl saw it divested of its feathers, they would find that the head, instead of being like a turnip, is no larger than that of a chicken, and very similar in shape. It is the feathers only that make it appear so large and strange in its formation. Were I a farmer, I should be as particular about my owls as any one thing on the farm. I should give every one in my employ notice, that if they wilfully injured or molested any of them, that instant they would be discharged; and as to strangers, I would prosecute them for trespass.

Whatever praises you may bestow on a good cat for in-door services, may with equal justice be bestowed upon the owl for out-door services; and whatever advantages a cat may have over the owl for killing mice in close quarters, are more than counterbalanced by his powers to pick them off the thatches, and pounce on them in the open grounds, where the cat never goes; and such are the advantages given him by the use of his wings, that he can scour an acre of ground, hedge-rows too, while the cat is watching at a single mouse-hole. But so similar are the two in appearance, habits, and manners, that to sum the owl up in little, we may call him the feathered or flying cat.

There is another peculiarity also with regard to their manner of breeding. The hen generally lays three or four eggs; but often, before the young ones are half fledged, she is sitting again on a fresh lot of eggs in the same nest. Consequently, if farmers only protected them, they would soon have owls in abundance; and, to use Mr. Waterton's words, they would repay them at least a hundred-fold for their trouble by the enormous numbers of mice, insects, and young rats they would nightly consume.

I have devoted some little time to the consideration of this subject, because I am satisfied that farmers in general have never given these birds that serious consideration they merit; or they never would have allowed such invaluable creatures to be treated as they have been. Farmers should know that owls, cats, and poisons are their only securities against those destructive little pests, the mouse and shrew tribes. But of the three, the owl is by far the most valuable. As to the use of traps for catching mice and shrews on a farm, 'tis somewhat like killing a fox by a hair at a time; for they increase much faster than you can catch them.

Now let us suppose a farmer to have only thirty barn owls upon a farm of one hundred and fifty acres. That would average one owl to every five acres. Then let us average them to catch but two mice each, nightly, the year round! and at the end of twelve months, pray what would be the amount of mice destroyed? Not less than 21,900! Let us further imagine these mice to be all living, and that ten of them will eat and waste one wine-glass of grain, at strike measure, in twenty-four hours, what would it take to keep the whole? No less than four bushels, two gallons, one pint, and three quarters, daily,—or ninety-seven quarters, four bushels, seven pints, and three glasses, in the half-year. But to say nothing of the odd pints and glasses, and supposing the mixed grain at fifty shillings per quarter, what then would be the cost of keeping these mice for six months? Why no less than £243. 15*s.*

I shall now conclude my remarks on the owl with two more observations. Suppose one mischievous bird out of the thirty did now and then destroy a little chicken (though how they are to do it I cannot tell); but suppose they did, —pray what would be the cost of the damage done? Would it amount to a sovereign in six months? I think not; but supposing it did,—well, then, the damage done to the farmer by thirty owls, in six months, would be one pound, while, during the same period, the property they would save to him, by the destruction of his vermin, would be £243. 15*s.* And all this is on the supposition that they destroy only two mice each in four-and-twenty hours.

The next consideration is regarding their doings among rats. In the first place, 'tis certain that rats, like every other animal, must be young before they can be old. In the second place, 'tis certain that owls like a young rat for breakfast as well as they do an old mouse, if not better, because it is more delicate, fat, and tender. In the third place, there is no doubt that owls, the same as ourselves, like to go where they are the best and quickest served. And in the fourth place, it is equally certain that young rats, like kittens and young rabbits, are very fond of playing and gambolling about the mouth of the hole as soon as they have strength enough to do so; and thus become an easy prey to the lynx-eyed owl. You may be sure that where the owl has caught one, he will not be long before he wends his course that way for a second; and so on, till he has devoured the whole litter; thereby not only preventing the possibility of their becoming old, but cutting off their entire posterity. There is no doubt, if farmers had a goodly number of these invaluable birds about their premises, that in a short time rats, mice, and shrews would become infinitely more scarce than they are now. Therefore, in conclusion, let me advise farmers to have holes cut in the sides of their barns and granaries for these birds to go in and out at, with footboards on each side for them to alight on, and also to have large boxes hung high up on the inside for them to roost and breed in. Then may farmers bid defiance to all kinds of vermin, after having first well cleansed their farm and premises themselves.

Now we come to the true friend of man, the remorseless destroyer of rats when he can get at them. He will protect his master night and day from thieves of every denomination. Nothing can shake his fidelity; and all he requires, in return, is a crust and a friendly pat. With these he feels amply rewarded, and will exert every faculty with which nature has endowed him to serve and amuse his master. I mean the well-bred Bull-Terrier dog. I say the bull-terrier, because the thorough-bred terrier, though an active, sagacious animal, and very fond of hunting, is nevertheless a very careful one, and kills a rat more by cunning than courage. He likes to wait his opportunity, and catch the rat while running, so as to give him a nip without having a bite in return. This you may say is sound generalship. So far, so good. But if there happen to be thirty rats present, twenty-nine will make their escape while he is dodging and fretting over one. Still I am satisfied, that if you take dogs in general you will not find more than one in fifty that will kill a rat; and if you lump all kinds of terriers together, both rough and smooth, I am equally satisfied, that, where you will find one that will kill ten rats off-hand, you will find ten that will not kill one each without the assistance of their master. They will do all the fretting and barking, if the master will do all the thumping and kicking; and thus kill the rat between them. But often, when the rat is dead, to the great delight of the master, the dog will give it a most unmerciful shaking, and thereby earn for himself not only a host of caresses, but a wonderful reputation.

When I was about nine years of age, my father bought me one of these highly reputed terrier champions. Toby was his name; and fifteen shillings was the sum paid for him. His exploits among rats were so startling and numerous, that I expected nothing else but that he would eat them alive. However, he had not been home long before all his powers were put to the test. Into the back premises we went, "Rat, rat! Toby, rat!" Out dashed an old fellow from behind some tubs, and hearing the dog's squeals, in his haste he popped into the wrong hole,—one that was only about a foot and a half long, and closed at the other end. Toby screamed, and tore like a raving mad lion, and set to work most furiously to tear up the very foundation of the place. Down came all the workmen, and cheered him on. Up came the bricks and stones one after the other, which he removed with his mouth, till in his eagerness and excitement I verily thought he would have swallowed some of them. At last, on removing a brick, out sprang the rat. Heels over head went the men, one over another. Away went Toby, ready to break his neck, out at the door, and never stopped to see if the rat was after him, till he reached the other end of the street. Then finding all was safe, he quietly sat himself on end in the middle of the road, and not all the whistlings and enticings possible could induce him to come back, till he was satisfied the rat was gone. However, his courage returned again, and he went through the same fury on the morrow. But when my father came to see the place, he sent master Toby about his business. At the same time he sent for the bricklayer to put the place in repair, declaring that the dog had done more mischief in a few hours than the rats had in as many years. And thus I lost my hero!

Since then I have, at various times, had at least half a hundred terriers of one sort or another, but there was only one out of the whole that would kill a full-grown rat single-handed; but even he was very soon satisfied, since he mostly declined killing a second till another day; and this I have found to be the case with the great majority of thorough-bred terriers. The truth is, they are too cunning and too soft for such hard work. But when they are bred in with the bull-dog, then you have the most active, resolute, and hardy dog that can be produced; and all those dogs that have performed such wonderful feats in the art of rat-killing are of this breed.

The great object, among the various breeders of these dogs for rat-killing, is to have them as nearly thorough-bred bull as possible, but at the same time preserving all the outward appearance of the terrier as to size, shape, and colour. Black and tan are considered the essence of perfection. The head, neck, body, and tail must be jet black, and not the shadow of a white hair about them. The legs, feet, chest, under jaws, and glottis must be the colour of a deep, ripe chestnut, with a full round spot over each eye. The hair on every part must be very short, fine, close, and glossy; the feet long and extremely narrow, with long black claws, and a pencil mark or black streak up each toe; the head round, and firmly fixed on an arched or longish strong Roman neck, well set in the shoulders; thin, transparent ears, cut clean out at the bur, and brought to a graceful point; eyes black, bright, prominent, and well set; jaws full and firm, but rounding smoothly off to a muzzle of sufficient length, strength, and substance; small, thin lips; nose flat at the point, with inflating nostrils; fangs long, strong, and straight; chest deep and full, but not too broad; body rather short than long; loins firm, but gracefully working off to well-rounded haunches, rather light than heavy; and the whole must terminate with a thin, tapering tail, about the length, shape, and substance of a highly-bred young lady's delicate little finger. His action must be bold, yet graceful as an Arab steed's. At the same time he must be agile as a kitten, and as springy and elastic as an India-rubber ball; but in his every movement, look, and expression there must be an air of whining, restless, dauntless defiance. His weight should be from ten to fourteen pounds, not in starved, but trained muscular condition. With these requisites you will have a dog that may be pronounced a perfect specimen of a black-tan bull-terrier.

The Royal Rat-catcher dog, is certainly one of the most beautiful, and at the same time the most serviceable and interesting creatures in the whole animal creation, and well worthy the attention and caresses of a right royal master. He is fit to be sent as a royal present from monarch to monarch, or to grace the refulgent equipage of hereditary sovereignty. For as we have the grand falconer, and masters of the horse and hounds, why should there not be the royal rat-catcher, dressed in top-boots or patent-leather leggings, buckskin breeches, a silver-mounted green velvet coat, with a gold-embroidered rat in front of his hat, a silver rat-cage at his back, and three or four such dogs in the leash, secured with silver-mounted muzzles? Pray, would not this form an interesting addition to the royal procession? or would it not give as great an impetus to the downfall of vermin, as was given, on the contrary, to the red-herring trade by his His Gracious Majesty, George III., having a dish of those fish brought daily to table?

I will leave this matter for the consideration of persons of rank and influence. But if such a man were chosen, he should be one well qualified to deliver interesting lectures through the country upon the evil effects of vermin, not only to individuals, but to the nation at large.

Some have asserted that there never was such a perfect specimen of a dog as the one I have described; but that is a mistake, for the model I chose was a dog by the name of Pincher, and the property of a gentleman, who bought it of a dyer, in Oxford. Pincher had been my property; but he had been ill-used and shyed when a puppy. Still, for exquisite symmetry and brilliancy of colour, he was the most perfect specimen of a black-tan bull-terrier of any man in England.

Thus we perceive that the dog-fancier is not of that low, ignoble character that some people would wish to rate it; since we find among them, not only humble artisans, but kings, queens, princes, and princesses, in every period. Ladies of all ages and degrees, lords of every grade, and priests, parsons, and poets. Indeed show me, if you can, a fancy that embodies such a host of talent of all ages, ranks, and nations, as the dog fancy. Still I admit that all do not select black-tan bull-terriers for parlour boarders or nursery companions; yet each has his fancy, and let him enjoy it, so long as he does not abuse it. At the same time bear in mind that there is a vast difference detween dog-fanciers and dog-stealers,—just about as much as there is between the man robbing and the man robbed.

The scoundrel dog-stealers spoil the appearance of nearly all the best dogs in London, because, to prevent their being stolen, their masters are compelled to keep them constantly chained up, from their puppyhood upwards; and by their being continually set on their haunches, their fore legs become bowed, and their hind ones contracted, which entirely spoils their grace and action. Again, many breed these dogs as small as possible, and then stint their growth by constant starvation. Herein there is great barbarity practised for the sake of fancy and fashion. But many of these little black-tan pigmies, known as toy dogs, are not bull-terriers at all, but greyhound terriers, having been bred from the Italian greyhound; and, for any service in defending their owners from rats, they are not worth a pinch of salt. So much for fashion and fancy.

True sportsmen and rat-catchers tell us that a good dog cannot be of a bad colour; that is to say, it matters little what the colour may be, so long as the animal is a good one. A dog for sound service should be any weight between eight and eighteen pounds. But one from ten to twelve pounds is a most serviceable creature, and, if well trained, will work its way through almost any number of rats. A well-trained dog never shakes a rat, or bites it twice, but seizing it across the neck and shoulders, pins it to the ground till it is dead. In this way many of them will kill from ten to twenty rats in one minute, if in close quarters. Nor do they ever bark or scratch at a hole, unless urged to do so, but patiently wait where you place them till the ferret is put in, and the rats bolt; then they do their work as quickly as possible, and quietly wait for more.

Bull-terriers also vary very much both as to courage and quality. Even puppies of the same litter, though all taught alike, will often differ as much as so many dogs of different breeds; and sometimes not one among them will turn out worth a rush for rat-kiiling. Still there is this to be said, that if you have a well-bred puppy, you stand a good chance, by careful training, of having a decent killer, perhaps a good one; while, on the contrary, a mongrel, with all the training possible, never can make a good dog. But I must tell you, that an ill-tempered, surly, vicious man never yet brought up a good dog. You may give him the best puppies in the world, and his coarse, morose temper will spoil every one he has to do with. So you may take my word for it, that a kind, gentle, cheerful disposition is the first and principal qualification for a good sportsman, or trainer of young dogs.

Few of the best rat-killing dogs of London understand anything about rat-hunting, and would as soon kill the ferret as look at it, never having been trained to them; consequently, though some of them would kill any number of rats in a pit or parlour, if properly secured from escape, yet at the same time they would kill scarcely a rat a month of their own catching.

Perhaps the most extraordinary animals for rat-killing that the world ever saw, were two celebrated dogs, named Tiny and Jem, of whom I have already spoken at some length. A small account of these most wonderful creatures may not be wholly uninteresting; and therefore I will wind up the subject on dogs with "Tiny and the Baby."

Tiny was a very slender, pretty black-tan bull-terrier, about the size of an ordinary cat, and weighed only five pounds and a half in trained condition. He was a most excitable little creature, and could not stand still for one instant. When his master brought him into the parlour to show me, I certainly never saw such a sight. I could scarce tell which was the head and which the tail; for he went round about, up and down, in and out, this way and that, with such rapidity that I could form no idea at all as to his size, shape, or colour, except that he looked like an India-rubber ball with glistening red streaks about it; or rather like a bundle of affrighted eels twisting and twining in and out of each other for the purpose of hiding. But as he grew older, he became more steady and dignified, and used to sit in state, on a crimson velvet cushion fringed with gold lace, and placed on the bar-parlour mahogany table, with large bright candlesticks and mould candles on each side, so that visitors might see him from the front of the bar. And this was Tiny, the then rat-killing wonder of the world, and conqueror in about five thousand life and death battles with rats.

I shall now give some of his wonderful performances.

When nine months old, and weighing only four pounds and a half, he won two matches at six rats each. When he weighed five pounds, he won twenty different matches at twelve rats each, and fifteen matches at twenty rats each. His next match was to kill fifty rats before he was taken out of the pit, which he won cleverly, never stopping till he had destroyed the whole. Tiny was then matched to destroy one hundred rats in thirty-five minutes, which task he accomplished in thirty-four minutes and fifty seconds, winning the match by ten seconds. He afterwards defeated the celebrated Somer's-town dog, Crack, eight pounds weight, in a match at twelve rats each. He beat the renowned dog, Twig, six and a half pounds weight, at thirty rats each; and was matched to kill two hundred fair barn rats in three hours. This task he accomplished in fifty-four minutes and fifty seconds! thereby winning the match, with two hours, five minutes, and ten seconds to spare. Tiny was matched to destroy twenty of the largest rats that could be produced, in ten minutes, and which task he completed in eight; thereby winning the match, with two minutes to spare. A fortnight afterwards he beat the celebrated bitch, Fan, eight pounds weight, in a match at fifty rats each. He was then matched to destroy fifty rats in twenty minutes, without any one being in the pit with him, and which task he accomplished in fifteen minutes and twenty-five seconds. He was matched to kill twelve of the largest rats they could find, in three minutes; which he won in two minutes and thirty seconds.

We next find him struggling by himself with fifty large rats, having been backed to destroy them in twenty-one minutes, but which he did in twenty minutes and ten seconds. A month afterwards he was backed to destroy one hundred rats in thirty-one minutes, and which he completed in thirty minutes and three seconds. Again, he was backed to kill another hundred in half an hour, and which he accomplished in twenty-eight minutes and five seconds.

Tiny was again pitted with two hundred rats. This match was a close run, having been backed to destroy them in one hour, and it took him fifty-nine minutes and fifty-eight seconds; thereby, winning the match by only two seconds. He was again backed to kill one hundred fair barn rats in half an hour. This match he won in twenty-nine minutes and ten seconds. He was backed to kill twenty of the largest rats they could find, in four minutes, which feat he accomplished in three minutes and seven seconds; and a few days afterwards he was again pitted with twenty of the largest rats they could produce, being again backed to destroy them in four minutes, which he completed in three minutes and fifty seconds.

To say nothing of private matches among gentlemen,

Tiny, when he died, had contended in upwards of fifty public matches, all of which he won. Suffice it to say, he had never suffered a defeat, and had destroyed over five thousand, or nearly a ton and a half weight of rats. He had also been presented with numerous beautiful and valuable collars, both by gentlemen of rank and public subscription. On one occasion a nobleman offered a hundred pounds for Tiny; or rather, he offered a hundred guineas for Tiny and a lesser dog, worth only five pounds, which left a hundred pounds for Tiny; but which offer was promptly refused by his master. His manner of killing was different from anything of the kind I ever saw. In a heavy match, when put into the pit, he would set as steadily to work as any little old man going on a journey. The rats always get into the corners, and there form pyramids, five and six layers deep, resting on each other's backs. Sometimes, if they do not stir them about, the under ones will be suffocated. But instead of dashing in among them, as most high-bred dogs do, Tiny would stand quietly at one heap, and pick out the largest first, and then be off to the next heap; and so on, till he had disposed of all the biggest. He never bit a rat twice, or ever shook one; but after he had dropped it, it was a pound to a penny that it never rose again. When he became tired, he would leave off, and lie down in the middle of the pit for his master to wash his mouth, and refresh him by blowing on him; but as soon as he had gathered his wind a little, he would up and at it again; and so on until every rat lay dead. On one occasion, all his front teeth fell out, except one fang, with which he finished the match victoriously. Several gentlemen begged a tooth each as a great favour, and had them mounted in silver and gold, to preserve as relics of this most wonderful little creature.

Tiny died from over excitement. Some men had a rat in the parlour, and though Tiny was chained in the bar, and could not see it, still, such was the state he had worked himself into, that they became alarmed; and though they let him kill it, he died soon after. On examination it was found that he had burst his heart in three places; at least so his master informs me. And this was the end of Tiny, the rat-killing wonder of the world. He was afterwards stuffed, and is now exhibited in a glass case. But his master tells me that for him as he is he would not take a hundred sovereigns.

Jem, the champion, was a fallow-coloured bull-terrier, about eighteen pounds weight, with a head nearly all white, and in his general appearance as plain a looking dog as you would wish to see, except that he had an unusually long, strong, square muzzle. But for steady perseverance and powers of execution he has never been equalled. His public exploits were numerous. He contended in eighty public matches: namely, 20 matches at 20 rats each, 30 matches at 50 rats each, 28 matches at 100 rats each, and 2 matches at 200 rats each; thus destroying in public 5,100 rats. The longest time he took to destroy a single hundred was eleven minutes and twenty seconds; and the shortest was five minutes and fifty seconds. This is the quickest time in which one hundred sound rats were ever fairly destroyed by a single dog. But to add to the wonder, Jem, when he had had but ten minutes' rest, was again pitted with a second hundred, and in six minutes and one second every one lay dead; thus destroying two hundred fair barn rats in the short time of eleven minutes and fifty-one seconds, or at the rate of seventeen each minute. This I believe to be the greatest feat in rat-killing ever performed by a dog.

The celebrated rat-killer, Billy, who exhibited some thirty years since, did not perform anything near the feat of Jem; for though Billy's time, in destroying a hundred rats, is stated to have been five minutes and a half, still, let it be borne in mind, and I assert it on the testimony of living witnesses, that numbers of the rats were dead before the dog commenced, and that the whole had been poisoned with nux-vomica before being put into the pit. This is the poison that rat-catchers give those rats that may sometimes be seen crawling about them in the streets. Of course they give them but little, or they would die too soon. It has the effect of partially or wholly paralysing them, according to the quantity they have eaten; and this is the supposed charm that many rat-catchers have over rats to tame them.

A gentleman, a friend of mine, who witnessed Billy's feat, leant over and picked up two or three of the rats that were crawling about, and he declares they were perfectly harmless, and not able to see. Not only that, but the instant the dog touched them, they were taken by the tail and slung out of the pit, whether they were dead or not. But as it has been shrewdly remarked, had Billy stopped at home till the following day, the rats would have saved him the trouble, because they would all have been dead before morning.

In our day there are different rules and regulations for rat-matches; for if there be one suspicious rat put into the pit, it is instantly replaced by another, so that they are all approved rats. Then, when the dog has done his work, and his master, or second, has picked him up, should there be any rats lingering, they are placed in the centre of the pit; and if they can induce them to rise and crawl the length of their bodies, the dog has to come in a second time, and finish them; and the time thus employed is added to the rest, thus does many a furious animal lose the match. And had Billy been bound by such rules as these, I am persuaded he would have been nearer twelve minutes than five in accomplishing the task. The truth appears to be that Billy did not destroy a hundred rats at all, for numbers were dead from poison before he commenced; and when he did commence, several were thrown out as dead that were able to crawl away. So that, taking these matters into consideration, the question is, did Billy in reality kill three-score rats in five minutes and a half, instead of five score? Not so with Tiny and Jem. They did their work to the satisfaction of every one present, both winners and losers, and never left a doubt upon the question.

The dog Tiny never suffered a defeat, but, on the contrary, Jem was twice defeated, if defeats they might be termed. In the first place, he was chained up for about three years, and not let loose except for a few minutes, for weeks together; and then only to kill a few rats, after which he was instantly chained up again. All at once he was matched against the greatest champion of the day,—Crib, of Richmond, in Surrey,—a highly-trained dog; and so prized by his master, that he spared neither pains nor expense to keep him in good condition. The match was to come off in a fortnight from the time of signing articles. A person was appointed to take Jem in hand, and see what could be done, by way of training, in so short a period. He chained him behind a vehicle for two or three days;

and the dog's feet (being naturally tender from inactivity, and the distance and the force with which he dragged him over the hard roads) were completely excoriated. Hence the poor creature had to run on the bare flesh. Added to this, the dirt and gravel were so completely ground in, that on the day of the match, instead of being, like his opponent, in first-rate condition, he could not stand, and was forced to eat and drink lying down. However, the match was play or pay, and Jem's master, not being willing to lose without a struggle, set his wits to work, and hence the following stratagem. He procured a number of babies' socks, and having filled them with softened fuller's earth and oil, thrust the dog's feet into them, and carefully tied them round with garters; then, when time was called, Jem was carried to the pit, and the instant he saw the rats, his courage overcame every obstacle, and, to the astonishment of every one, he set to work most gallantly, and never stopped till he had finished his task. But the pain and stiffness, and the entire want of all training, retarded his buoyancy and activity; and hence he was behind time with his powerful opponent. But another match was instantly made, to come off in two weeks, when Master Crib had to sustain a most inglorious defeat, not having the shadow of a chance. Thus was proved the then sounder-footed Jem's complete superiority over him as a rat-killer.

His second defeat, if it could be so called, was, if possible, more unfair than the first; for he was taken, as it were, off the chain, and backed against a full-trained dog, to kill a hundred rats to his opponent's fifty, thus destroying at the rate of two to the otner's one; but so near did he run the struggle, that no odds in money, not even a hundred to fifty, could induce his opponents to renew the match on the same conditions. In every other match he came off victorious, and was, at the time of his death, the undisputed champion of the world.

The dog Jem was the envy of the whole rat-killing fraternity, and such was the care necessary, for fear of foul play (I mean theft and poison, from both of which his master had been more than once a sufferer), that Jem was forced to be kept in almost constant seclusion and confinement, and never went out for an airing except with his master, when business would permit. On the last occasion he took him as far as Chelsea ; and when returning through Brompton he mounted an omnibus, for the purpose of giving him a run. The poor dog was following alongside, and looking up at his master, when a rival buss shot by, and, running over him, killed him on the spot. The grief occasioned in the district, as soon as it was known what had happened, is more easily conceived than described ; but suffice it to say, that a gentleman, to soothe the sorrows of his disconsolate master, presented him with a magnificent signet ring, with a miniature likeness of Jem in the act of killing rats, and encircled with the motto, "Every man has his fancy," most exquisitely engraved on the seal. The engraving cost eight guineas, and the ring eight more, which ring Jem's master wears in honour of the gentleman and his poor dog.

That these dogs are not the properties of the humbler classes, the following will prove most clearly. Jem's owner informs me that it cost him scores and scores of pounds to bring Tiny and Jem to their state of perfection. This I can easily believe, from the fact that they knew so well the difference between the words, *head* and *dead;* for if they seized a rat by the hind quarters they were almost certain to be punished for their trouble. To avoid this, their master would sing out *head,* which summons was always responded to by seizing the next rat across the head, neck, and shoulders. Then, when he sang out *dead,* the rat was instantly dropped, and another as quickly seized ; thus showing a great degree of perseverance and practice. But he tells me that Jem had destroyed over 10,000 rats. Here, then, can we calculate pretty nearly the cost of training him. In public he had killed exactly 5,100 rats, and that leaves about the same number for training. Now 5,100 rats, at prime cost, namely, three shillings per dozen, amounts to £63 15s. 0d. But if an amateur purchases rats for the purpose, he has to pay the retail price, which is sixpence each, or six shillings per dozen. In that case the training of Jem would have cost £127 10s. Now, whether the rats be bought wholesale or retail, pray what man in humble circumstances can afford the expense? No, the great majority of real rat-match dogs are the properties of persons who can not only pay the cost of training, but back them besides; and among which persons we may rank, not only publicans, but noblemen and gentlemen, both civil and military, as well as citizens of London and first-class tradesmen in Bond-street, Oxford-street, and Regent-street; besides master butchers, bakers, milkmen, and a host of others. Still, let it not be supposed that gentlemen either train or second their dogs in such cases. They can always hire persons for such offices, while they themselves can look on as casual observers. Nor am I aware that they are less qualified to fill their various positions in life, because they feel a fancy for dogs, or an interest in the destruction of vermin. But, after all, I must confess that rat-matches seem a good deal like hunting a bagged fox, or a Cockney sportsman filling his pockets with poultry in a farm-yard, instead of traversing over moor and mountain in quest of game. And though it is true, that in a pit Jem has destroyed two hundred rats under twelve minutes, yet, if left to hunt them in their natural runs and retreats, pray would he have killed two hundred in twelve months? or would he have caught as many rats in the year as one of those little flat-headed cats I have already spoken of? Remember, rat-matching is one thing, and rat-catching another.

But before leaving the subject, let me put my country friends on their guard against a most dishonourable practice, by which they are not only deceived, but lose their money; and that is, the practice of painting first-rate dogs, and matching them as novices. If I tell you of one celebrated animal, upon which this was often practised, that will be as good as a thousand, so far as the powers of painting are concerned.

A boxing-glove maker (of London) had a fallow and white bitch, named Rose, about fourteen or fifteen pounds weight, which was the fastest killer in England. This man told me, in the presence of a friend, that the late ex-champion was her principal backer; that he himself had not sufficient means, but that his backer always paid him handsomely for the use of her, and had won a deal of money by her both in town and country. It did not signify who had a dog near her weight, he always had a novice on hand, on the which he would risk his money, if they would stake a good round wager. Sometimes his novice was a red and white bitch; at others, a black and white one. Then it would have a black or red patch over one eye; or, if necessary, over both eyes. On other occasions she would be all one colour; and lastly, to drown all suspicion, his novice would be black and tan; indeed, she had been painted all the natural colours you could mention, excepting white. Thus were numbers fleeced of their money; for while they thought they had a novice to deal with, they were contending with the fastest killer of the age under the disguise of hair-dyes.

Take this case then as a warning; and if gentlemen will match their dogs, let them know against whom and what, or they may rue their folly.

I shall here relate a curious anecdote of "Tiny and the Baby."

About three years ago, when travelling by rail from London to Winchester, I saw a man who was waiting at one of the stations with a large wire cage, in which were about a hundred rats. He had all the appearance of a rat-catcher. The circumstance excited great interest among the ladies and gentlemen in the carriage where I was seated; and this led to the recital of many anecdotes of rats, from which I selected two of the most interesting for my note-book.

The narrator, it appeared, was an independent gentleman, living in the neighbourhood of Leeds, who was accompanying his wife (an amiable-looking lady) and their infant son, about three months old, to Southampton. But the hero of the party was a pretty little black-tan terrier dog, scarcely as big as an ordinary cat. He was seated on his mistress's lap alongside the sleeping baby, but evidently very unhappy; for he eyed with keenest suspicion every one that moved, and two or three times, when a lady attempted to touch the baby, he turned up his lips, and showed his little white teeth, which made her glad to desist, though it made her and every one else laugh.

The gentleman told the history of Tiny, for that was the hero's name, with the greatest joy and delight, though at the expense of a deal of suppressed laughter on the part of his good lady, which seemed to give her lord and master the utmost satisfaction. It appeared they had been only some twelve months married. About six months since, an old dog, that had been the companion of her girlish days, had died from extreme age; and the lady took the matter so much to heart, that her husband became alarmed. In addition to that, he found the country seat they had just occupied terribly infested with rats, and knowing her extreme horror and disgust of these animals, he became seriously concerned, not only for his embryo son and heir, but more especially for the safety of his amiable partner, who, from the natural delicacy of her constitution, and the precariousness of her then situation, was rendered infinitely more nervous and susceptible of the smallest fright or alarm than she otherwise would be. Taking all matters into consideration, he resolved upon buying her another dog, a pretty little pet, but not a spaniel, because (to use his own words) the one that died used to look more like a dirty old door-mat than a dog; and besides that, the breed are perfectly useless. No, he would buy her a handsome little terrier—one that would kill rats; then it would protect her, and perhaps do away with her alarm at these things! Besides that, he had frequently read in the sporting journal, of the wonderful exploits of the celebrated Tiny; and having some monetary matters to transact in the City, he determined upon starting at once for the metropolis, and if possible procure one of the little wonder's puppies. He arrived in London, and saw a dog-fancier. The result was the purchase of the present little hero, which at that time was about twelve months old, and named after its sire, Tiny. With all haste and delight the gentleman took it home, and introduced it to his lady, who upon first sight declared it to be an ugly little beast, and wished it instantly to be taken away; for it looked as if it had never had a meal in its life. This was a chilling reception, after so much trouble. But the poor little creature was cold and hungry, and in a strange place, and had just come off a long, noisy journey. It looked frightened and timid, its head and ears were down, its back cocked up, and it trembled all over; indeed, it was seen altogether to disadvantage. As to its slender appearance, that was its natural formation. Nevertheless, the lady's repulse had thrown such a blank over the whole affair, that the gentleman began to think he had been running a wild-goose chase after a bad bargain. For some minutes a dead silence presided over the tea-table. There sat the poor dog like a little criminal just taken in the fact. There was its master in profound thought, sipping his tea, and vacantly looking at nothing; and there sat the lady anxiously looking first at one and then at the other alternately. At last she broke silence by inquiring where he got it! "In London," was the prompt reply; "and if he had for one moment thought she could have felt so displeased with it, he never would have brought it home." She said it was such a little thing. "That, my dear, is one of its greatest beauties; and furthermore he asked her if she was aware that it was one of the most valuable and finest bred dogs in the country, and that a hundred pounds had been offered by a nobleman for its sire, and refused. You may well feel astonished, my dear, said he, but I have seen things since I saw you last, which, had I not seen, I could never have believed. Now, would you think it possible, that that little creature could kill a rat?" "Kill a rat," she exclaimed, with astonishment! "Yes, my dear, kill a rat. Then give him five more, and he will kill them also, and six more at the back of them. After that, give him another dozen, and he will not leave one of them alive." The lady looked perfectly bewildered; when he concluded by telling her that last evening he saw its father, for a round wager, destroy two hundred rats in less than one hour.

This information entirely altered the lady's estimation of the young Tiny. She at once saw in him all the marks of a warrior and a hero. Tiny was instantly brought to the fire, and petted up with warm milk in a saucer, on the hearth-rug, and chewed bread and butter and sweet biscuits; in fact, nothing now was too good for the little champion. She asked her then smiling husband how much he gave for it? and he whispered in her ear, "Jockeys never tell—I bought it for you, love; and what it cost is no one's business but mine."

As soon as the dog was warmed and fed, he began to fondle about his mistress; and by his little winning ways stole so imperceptibly upon her affections, that in a few minutes he was seated in her lap, licking her hands, and striving all he could to lick her face. This showed a degree of grateful attachment that was irresistible, and which at once established him the pet of the household. The gentleman was so pleased with the change, that he rose from his seat, and asked her what she thought of it then? at the same time accompanying his remark with a gentle slap on the shoulder, when the dog flew at him like a little lion! This, of course, was delightful! What! protect his mistress, even from her husband? Why, he was the dearest little creature in the world, and was cuddled up most affectionately. The gentleman continued teasing him by tapping his mistress, till at last, had he not jumped with his knees among the tea-things, the dog would have had hold of him. However, to turn its attention from him, he sang out "Rat, rat!" upon which the little creature hunted the room all over, and at last became so excited, that they picked him up, and made him again lie down in his mistress's lap.

The following morning, after breakfast, the lady not feeling over well, a ride out in the chaise was proposed; and Tiny should go with them. While she got ready, the gentleman would go and see the horse put into the chaise, and settle other little matters; but scarcely had he left, and summoned the servants around him, than they were alarmed by a terrible squealing of rats, followed by a horrifying screech and slam of a door, and in an instant all was dead silence. The gentleman stood aghast, as if riveted to the spot; the servants standing round, stared him full in the face, with their eyes and mouths wide open. His hat fell from his head; but how, he knew not. Presently a great fat servant-girl, recovering her breath a little, said, "Lorks, sir, that's missus; an' the rats have gone an' kill'd her;" and then burst out roaring, "Oh, my poor dear missus! Oh, poor missus!" Thus far his apprehensions were verified. He turned from the spot more dead than alive, and, slowly entering the house, ascended the stairs, step by step, followed by his breathless servants, eagerly listening; but not a sound was heard, save their own footsteps. On arriving at the summit, he, in the utmost anguish of soul, placed his hands before his uplifted eyes, being unable to behold the horrid wreck, and thrust open the door, when "hush, hush!" fell upon his ear. He gave a phrensied start; his eyes rolled; he gasped eagerly for breath; but in every other respect he might have been a marble statue, so frozen was his aspect. What did he behold? Was it the idol of his heart a prostrate and lifeless corpse? No; what then was it? why, his pretty, amiable, nervous wife upon her hands and knees, underneath the table, seconding little Tiny in a life and death battle with an enormous rat!! Instantly a thought flashed across his mind, that she had gone raving mad! Down stairs he flew, upsetting the fat servant, who went rolling to the bottom. Out of the house he rushed, and ran along the road, with all his might, to fetch the old doctor.

The facts of the case were these:—The lady, feeling unwell, was seated listlessly by the fire-side, with Tiny in her lap. The gentleman had proposed that they should go out for a drive, and had left the room but a few minutes, when in bounced a female rat, followed by two males, a larger and a smaller. The big fellow seized the smaller one by the ear, and gave him a sound shake, which made the young gentleman give a most unmerciful squeal. The lady in her fright gave a screech, and jumped up with the dog in her arms; but instead of running out, she slammed the door to,—a thing very natural with ladies, when alarmed by the sudden appearance of intruders. Still, in this case, instead of shutting the intruders out, she had fastened them in, when Tiny became so infuriated and ungovernable that he struggled out of her arms, and down he went. The she-rat was running across the hearth-rug, when he laid her dead. He dashed at the next, and laid him dead by the sofa; and as the largest fellow of the three was running under the table, Tiny pinned him. But he was too large and too tough for the compass of the dog's mouth. Still he stuck to his work like a lion, and held tight to him. The struggle was a desperate one. They rolled over and over two or three times; and, though severely bitten, Tiny was not to be worsted. This gave his mistress courage, and without thinking of her situation, she went upon her hands and knees, to cheer and pat him on to victory; and just as Tiny had vanquished the rat, in came the gentleman, and hence his extraordinary suspicion as to the lady's sanity of mind.

We will now leave the lady, and follow her husband, who

had run out of the house at the top of his speed, without his hat, to fetch the doctor—a corpulent gentleman, about sixty years of age, who resided about half a mile off. On arriving at his residence, he rushed into the shop like a madman. "My dear sir," said the medical gentleman, who happened to be present in his morning gown, and without his wig, "what on earth is the matter?" "Matter," gasped the affrighted husband; "my wife, my wife,—mad!" With that he seized the doctor by the collar, and dragged him through the streets, to the great astonishment of all beholders. Every intreaty for time and patience was fruitless! On he dragged him, followed by a motley group of gaping spectators, till they were within two hundred yards of home. There they were brought to a standstill. But in vain did the asthmatical old gentleman expostulate against the impropriety of running him at such a rate! The husband began pulling and hauling him again; when the old gentleman succeeded in grasping a gate-post, and clinging tight to it. "My dear sir," said he, "pray do consider my age, weight, and constitution! And besides that, sir, what will my patients think of seeing me without either hat, coat, or wig?" All was to no purpose; for the gentleman pulled, and tugged at him till he was compelled to leave the post! At last the old gentleman came to the conclusion, that the best way was to get there as quickly as possible, and a few minutes brought them to the lady's drawing-room, where she was seated, bathing the dog's wounds. The husband, without any ceremony, burst open the door, and dragged in the doctor, who threw himself into an arm-chair. "There, sir," said he—"there is my poor maniac of a wife!" "Maniac, Charles?" said the lady. "Yes, maniac!" he roared. "I don't understand you," she replied. "Why, are you not mad?" he exclaimed. "Mad!" she reiterated, "why, what do you mean?" "No," said the doctor, who by this time had recovered his breath a little; "and were I consulted professionally to decide which was the maddest of the three, I should not be long in coming to a conclusion." "My dearest love," said the husband doubtingly, "are you not mad?"—"Mad, Charles; no, certainly not!"—"Then I most humbly thank God for His mercy!" and turning to the doctor, "Sir," said he, "from my inmost soul I beg of you ten thousand pardons, and offer any atonement that is in my power to make." "Then, sir," said the doctor, "have the kindness to give me a good stiff glass of brandy-and-water, to replenish my exhausted strength, and we will cry quits." A large tumbler of brandy-and-water was instantly supplied him, which in one draught he as speedily disposed off; and after gaping for a moment, then smacking his lips, he declared that that was good, but the next would be better! The glass was quickly replenished, when, after putting it to his lips a second time, and drawing it within an inch of the bottom, he declared himself more composed, and somewhat better. "But, my dear lady," said he, pointing to the old rat beneath the table, "pray what is that?" Here the lady gave a most vivid and energetic description of Tiny's *rencontre* with the rats! To have heard her denounce mankind as a race of cowards, and eulogise the dog's courage, one might have thought, that had the hearts of all the naval and military officers of Great Britain been pounded into one animal, they could not have made such a hero as her noble little Tiny.

All this seemed to give the old gentleman the utmost satisfaction, who was looking at the whole transaction, not as a matter of curiosity, but with the practical eye of a medical man. When she had done, the husband gave a description of his journey for the doctor; and then the doctor, to the great amusement of all, gave a practical description of their journey back, and after most immoderate laughter on all sides, the doctor wound up the matter by telling the gentleman that he ought to value that dog above all creatures next to his wife. For in his mind there was not the shadow of a doubt but that the resolution of the little champion, at the critical moment, had so completely inspired her with that degree of nerve and courage, which had so happily frustrated the awful consequences of so serious a fright, in her delicate and peculiar situation, and thereby in all probability robbed, not only the grave of a mother and her child, but the gentleman also of a straight-jacket in a lunatic asylum.

Here let me wind up the subject of rat-killing dogs with two or three remarks. The fact of the lady's being, as it were, inspired at the moment with a kind of superhuman courage, through the resolution of her pet dog, is a circumstance of by no means rare occurrence. I have known many cases of the kind not only among ladies but gentlemen also; for when does a man feel more secure from unwelcome intruders than when he has his faithful friend and guardian, his dog, by his side? Then why should not ladies have their pretty little terrier as a pet and protector, and thereby avoid the often miserable consequences resulting from fright by mice or rats?

After having estimated rat-match dogs and rat-pits at their highest possible value, let us, in turn, bring them to their proper level. In the first place, we must bear in mind, that all the rats are first caught, and then carried some miles to the pits, for these dogs to kill. Consequently these dogs are neither more nor less than rat butchers, and the pits but mere rat slaughter-houses. Secondly, there is this fact staring us in the face, which says, would it not be far easier for us, when we have the rats secure in cages, just to drop them, and the cages too, in the first pond or brook we come to, and thus not only save these dogs the trouble, but render both them and the pits entirely useless? And lastly, if farmers in general will only do their duty in following out the plans I have laid down, not only will they enrich themselves, but increase national independence, by doing away with the necessity for the importation of so much foreign corn, while, at the same time, rat-pitting will die a natural death, from the want of rats to carry out their matches. But unless farmers bestir themselves, and do their duty in this respect, it will remain a matter for serious consideration as to the propriety of putting down rat-pitting,—a system which, from London alone, annually lets loose above 100,000 rats, with all their hungry and multiplying progeny, to feed and fatten upon the produce of the land.

Before closing this chapter on the "Police of Nature," I shall briefly notice the physiology of perspiration and of instinct, by which animals are guided in their propensities and natural pursuits; more especially as this knowledge is necessary for the proper understanding of the plans I am going to lay down in the two next chapters, with regard to the baiting of traps and the mixing of poisons.

In the first place, perspiration is that compound fluid which passes out through the pores of our skins, commonly known as *sweat*, and which sweat is composed of carbonic acid gas, human fat, water, &c., and is not only most offensive in its odour, but to a certain extent poisonous.

In the next place, perspiration is divided, as it were, into two sorts, namely, the sensible and the insensible perspiration. The sensible perspiration is that moist state of the skin occasioned by heat or exertion, and which we can both see and feel, and wipe off if necessary; but if viewed through a very powerful magnifying glass, it will be found spurting a foot or two from every part of the body, like so many little water-spouts, and in such numbers, that the body on every side seems enveloped in steam.

The insensible perspiration is that gentle discharge from the pores, which is ever going on, and renders the skin soft, but not wet, and spurts only a few inches from the body; but at all times it is regulated by temperature, action, or excitement.

Now we can understand what we call a cold. If we get chilled in any way it stops up the pores; consequently, the carbonic acid gas and humours of the body are kept in, and then we feel feverish and unwell, and begin to cough. What is more, we shall not be better till exercise, or physic, or some other agency, has opened these valves, and let out the noxious gas, &c. Everything we wear, even to the very soles of our boots, is, I may say, saturated with perspiration. This not only shows us the necessity for clean linen, &c., but the great mischief also attending the ordinary use of water-proof clothing; for that solution of India-rubber, which so effectually keeps out the wet, does as effectually keep in the perspiration. At the same time we can perfectly understand how it is that different kinds of animals can track us out by scent.

The most striking instance of this kind I ever met with was reported in the papers about thirty years since. Some gamekeepers came to the spot where a deer had been killed in the night time. In order to discover the poacher, they put an old bloodhound on his trail. The dog tracked him to the neighbouring town, and though it was market-day, and all the people busy buying and selling, he scented him through the market-place to a little public-house in a narrow street; then went up stairs to a back garret, and upon the keeper's securing the dog, and opening the door, there lay the poacher in bed, and the deer on the floor. This at once shows how powerful must be the odour of the human body, arising out of its natural perspiration; or how could the dog have scented out the man's footsteps, and that too, not only so many hours after he had reached his home, but through a public market on a market-day?

CHAPTER XI.

TRAPPING OF RATS, AND THE VARIOUS KINDS OF RAT-TRAPS.

THE traps in most common use are the box-trap, the wire-trap, and the gin, or steel-trap, each of which is good, as far as it goes; that is to say, when properly baited. But they will catch only one at a time, unless by accident; for if the spring be fixed too hard, then a second or a third rat may get in before it goes off, if it goes off at all. However, generally speaking, they catch but one rat at a time, and for that purpose each is equally good, if properly managed. But, generally speaking, persons unskilled in the art manage their traps so badly, that rats as scrupulously avoid them as they would a strange dog or cat. Thus is the remark so often made, that rats are as crafty and cunning as we are, and know what a trap is as well as we do. But this is erroneous. Traps have been brought into disrepute not in reality through the wisdom of rats, but through the inexperience of those who set them. Few persons consider that the rat has the sense of smell as keen as the bloodhound, and by instinct feels that man is its enemy; consequently, if the hand of man has touched either trap or bait he instantly smells it, and will turn away.

Now the first principle of the profession is, to handle both trap and bait as little as possible. Indeed the bare hand should never touch either; and the same care is as necessary with the mouse-trap as the rat-trap. But generally speaking, when a youth makes up his mind to set a trap, he brings forward a dirty, filthy, rusty thing, and then becomes so excited, that the perspiration pumps out at every pore of his skin, and, by way of assisting it, he stoops down in front of a large fire to toast a piece of cheese, herring, or bacon, till it becomes a matter of great doubt as to which is most greasy of the two, the boy or the bacon. Then when it is done he seizes it in his sweaty hand, and after well pawing it over, by sticking it on and tearing it off the hook some half dozen times, all the time besmearing the trap with his other hand, it is pronounced fit to catch a rat! But if it did, it would astound the whole of the profession; for though we may look upon the rat-trap with so much loathing and disgust, still, I can tell you, that it is not more repulsive to our eyes than is the scent of our hands to the sensitive nose of the rat. Consequently, if we do not wish to spoil all our efforts, and suffer disappointment, we must treat the trap and bait with all the care and gentleness that we should the most delicate piece of mechanism; otherwise we destroy its efficacy.

Having described the wrong way of treating traps, allow me to put you in possession of the right method of managing them.

In the first place, let your trap be cleansed from every impurity with a brush and water, and then hung up or put in front of the fire to dry. When ready, put on a pair of thick gloves scented with two or three drops of the oil of anise seed. These gloves may be kept for the purpose. That done, then select your bait, which may be a piece of cheese, herring, liver, or bacon; no matter which, for all are equally good. Now toast it on a fork; but when you turn it, do it on a clean plate with a knife, and not your fingers. Then, when it is done, put one drop of the oil of anise seed, or caraway, upon a piece of writing-paper, and rub the bait on it, and that will produce the desired effect. But if you put more than one drop the thing is spoiled, because it smells rank, and instead of enticing the rat will drive him away. Some fry their baits; but I do not think it so good as toasting or broiling. When all is ready force it on the hook with the fork, thus never touching it with the hand. After that, see that it is properly set. If it be a gin-trap, you must force the bait under the string or wires as best you can ; then see that the slide is down, for many a rat has carried away the bait through this piece of carelessness. When thus far complete, place it where you like, but not close to the hole, and lay it down as if by accident ; for all kind of formality excites their suspicion. A corner is a good place, as they always scent round the walls first, and by having a little trouble to scent out the bait, they run more greedily at it. Now leave it ; and ten to one but you have a rat before morning, should one pass that way. Still, if unsuccessful, do not disturb the trap, but leave it where it is ; that will lull suspicion ; for some old rats are very crafty and cautious animals, while the younger are less suspicious, and more easily trapped. But after every rat caught, cleanse the trap as before ; or all is to no purpose.

Should you be troubled with a grizzly old fellow, that carefully avoids all your allurements, adopt the following plan. Send to the butcher for a bullock's thumb. He will understand what you mean ; it is a particular part of the liver. That done, you may either fry it or broil it, just as your fancy leads you ; but in either case let it be nicely cooked. When ready, fork it on to a clean plate, and after cutting off sufficient, scent it as I have already described, and bait with the same care. After that, set an old chair or stool against the wall, or in the corner ; and after placing the trap under it, so as to allow sufficient room for it to work easy, cover the seat of the chair or stool loosely over with long straw, so as to reach the ground on every side ; but mind there is not too much straw, and that it is quite clear of the trap. Then it is a thousand to one but you have him. This method of hiding the trap I have found most effectual, as the darkness and secrecy of the place takes him off his guard.

Here I have given you the grand secret of the profession ; and to any one troubled with rats, it is worth ten times the price given for this book.

I will now give you the real artistic method of disguising the natural odour of the hands ; and this does away with the necessity for gloves. It is done by getting a handful of oatmeal, and dropping into it three or four drops of the oil of anise seed, or caraway ; then mix them well together by rubbing them through your hands. This do continually, till the baiting and placing of your trap are complete. The meal absorbs the perspiration, and the oil imparts to the hands a most delightful and delicate perfume, which, so far from being offensive, is most grateful and enticing to the olfactory nerves of rats.

I shall here observe that there are a variety of devices, or rather traps, all of which doubtless are good, so far as they will hold a rat when he is inside; but how to get him there is the matter for consideration. Here, then, we shall find the art of managing and baiting traps to be a science quite as necessary as the traps themselves.

In addition, the following may be considered as an excellent mixture for drawing rats.

Take half an ounce of the chemical oil of lavender, a quarter of on ounce of the oil of rhodium, and one teaspoonful of the oil of rats; put them into a small bottle, and well shake them before using. Still I think either of the oils used separately is quite equal to mixing them.

And now, in conclusion, I shall append to this chapter some useful and practical information derived from "Whistling Joe, the rat-catcher," of whom some ample notices have been already given in Chapter XVIII. of the first Part of the volume.

"To make assurance doubly sure," says Joe, "I had a three-jointed fishing-rod, at the top of which I fastened about a yard of string with a herring tied by the tail. About two feet from that, I tied a calf's tail scented with the oil of anise seed. Two feet from that again I tied an old rag scented with the oil of caraway; and then, to prevent their scenting me, I spread a few drops of the oil of rats on the soles of my boots; but as soon as I arrived at the squire's, I threw the trails into the rick-yard for them to eat; and then, taking off my boots, ran for some distance in my stocking feet to break the track. But in all other cases (for instance, to draw them from one part of the house to another, or from their holes to the trap, or if I found any holes in a barn, &c., that led into orchards, fields, or what not), I trailed the herring round the premises to the hole or holes where the traps were laid, and mostly baited with the same. This plan I have found most successful; but I never failed to cover the traps loosely over with straw or fern, so as not to check the action of the trap, and at the same time to leave a small opening for them to enter secretly.

"There were two other points I always observed in setting and baiting traps. The first was to rub the ends and the insides at the entrance with the herring; or dipping a little piece of twisted-up paper into either of the scents I might fancy to use with any other bait, and touching the ends and sides, throw the paper in: this had the effect of throwing them off their guard, by causing them to smell and lick the parts, and then rush eagerly at the bait.

"The second point was, always to strew the inside and round about the traps with chaff or sawdust, more especially wire-traps; and as to gin or steel traps, I always covered them with it, because the iron striking cold to the rat's feet, causes him to jump back, and mostly to turn away with suspicion.

"I must now tell you how I first cleansed the squire's estate of vermin. It was, to trap and ferret out all I could, and then poison the remainder. But let me guard you against one prevailing folly in these cases, namely, impatience. Many think, because they have gone to the expense and trouble of buying and baiting a trap, that the rat is in duty bound to run instantly into it, and be killed. No such thing; the rat has no more desire to be killed than we have; and what is more, it is a hundred to one if he goes near it the first night; for in most cases an old rat must first become familiar both with the inside and outside of the trap, before you have a chance of taking him. Hence the necessity for never moving a trap after you have once placed it; and you may take my word for it, that a fractious, ill-tempered man will never make a good rat-catcher. I have known rats to go for weeks before they would venture into a trap. Others, again, may be caught the first or second night.

"When I went to clear a place, I always made a practice of fastening up the drops of the traps with pieces of stick, and feeding the traps also, for at least three days and nights, so that the rats might run in and out, and feed at pleasure. Indeed, I never attempted taking them till I found them enter the traps boldly, and devour the feed freely; because should they be startled at the commencement by the springing of a trap or two, they are taught to look with suspicion on them, and may not come near them again for weeks, if at all. But when they have been seasoned to them by running in and out for "the feed," they become so familiar, that so far from avoiding them they will run into them for safety; and thus may you catch rat after rat, till you have secured the whole. But when you are catching, and go the rounds, always have a small wire cage with you to run them into from the traps. This cage should have a fall at one end for them to run in at, and a door at the other, to empty them out into the store cage; or, if you please, you can drown them first, and then throw them out.

"There was another practice I had of drawing them, which was regulated according to the place I was employed at. If at a dog-kennel, I used to put some small pieces of boiled flesh inside the trap as well as the feed; if in a slaughter-house, then some pieces of fat or fatgut; if in a brew-house, or malthouse, then I would strew some of the best malt I could get; if at a still-house, or mill, then some meal; or in a barn, or granary, then some of the best corn, as well as the feed. In short, whatever food the place afforded them, the same I used to strew about, to lure them on to the feed. And now I will tell you how I made 'the feed.' Take one pound of good flour, three ounces of treacle, and six drops of the oil of caraways; put them all into a bowl, and stir them well together, till it looks all alike; then cut a pound of the crumb of bread into small pieces, and stir it up with the mixture. They like the bread mixed with their food better than the food alone, it being too luscious, for which reason they do not like it so well by itself; but that night on which you catch put no bread to it, lest they should carry it away.

"This was the only means I used for years for taking rats with the old hutch traps, and I have caught some thousands.

"I must tell you," says Joe, "that every rat-catcher you meet professes to possess some great secret, unknown to any one but himself, and thus it is that they wear such an air of mystery and self-sufficiency. But, in every case, their mysterious knowledge only extends to the mixing of some poisons in certain proportions, or the mixture of certain oils for enticing rats. A namesake of mine died some years since in Hertfordshire. He was a singular creature, for though he would cook rats, and eat them with the utmost relish, still he could not find stomach enough to extract the oil from them. However, he used a very good mixture for scenting traps; for which the following is the recipe:—

"'Take twenty drops of the oil of rhodium, seven grains of musk, and half an ounce of the oil of anise seed; put them into a small phial, and shake them well before using.' The reason for mixing them, he stated, was, 'that in some places rats love the scent of anise seed; in others, they like the odour of musk; then, again, in others, they delight in the scent of rhodium; so, by mixing them together, they could take their choice.'"

CHAPTER XII.

POISONING OF RATS.

There are various secrets for poisoning both rats and mice; but it appears that Whistling Joe, during his long professional career of half a century, only used one kind, and he tells us how he applied it. "In the first place," he says, "always buy your poison at a substantial wholesale druggist's, because you stand a greater chance of having it pure. There is no doubt but many have failed in their attempts, through the adulteration of the articles used. Having warned you thus far, now take three ounces of treacle; then put to it one ounce of finely-ground *nux-vomica;* and after well mixing them, put them into one pound of the best wheaten flour, and add six drops of the oil of carraways; then stir them well together, till they look all alike, the same as I told you in making the common feed; then mix it well up with a quarter of a pound of bread-crumb, cut to the size of peas. Here will you have bane enough to kill half a hundred rats, should they equally partake of it.

"For mice," says Joe, "I always used the same mixture

as for rats; but without the bread-crumb or scenting oils, because they do not seem to like the scent. In a dwelling-house, I always had the place cleared of all food; then spread the mixture on a sheet of paper, and in the morning, when I went to clear it away, there often would be numbers lying dead. This I repeated night after night, till there were no signs of any left. But I do not recommend poisoning mice in dwelling-houses, because of the danger of poison lying about: they are easily trapped; or if you procure a half-grown kitten, or one of the small cats spoken of, it will speedily clear the place."

Whistling Joe's poisonous draught for barns and ricks was as follows:—Take a quarter of a pound of the best powdered *nux-vomica;* put it into an old saucepan, with three quarts of water, and boil it till it comes down to two quarts; then put into it two pounds of treacle, to overcome the bitterness of the *nux-vomica;* when cold, pour some into little earthen pans, and put them in different places under the eaves of the ricks, and on the top of the walls and cross-beams in barns. Both rats and mice will greedily drink of it; and it will certainly kill them. But if you have owls in your barns, it will poison them also, if they drink of it.

In his concluding advice and instructions, Joe gives us some of the pretended secrets of other rat-catchers. "I have known it a practice with some," says he, "to resort to a variety of means, for the purpose of fumigating or frightening rats away. But that I consider merely trifling with the matter, because they only drive them on to the neighbour's premises for the time being; and before the week is passed they may be back again.

"Some rat-catchers," says he, "carry a pepper-box filled with finely-powdered arsenic; then getting some dead rats, they shake a quantity of the poison over their fore parts, and place them in the drains or holes of other rats. If they find these creatures have laid siege to a corn-rick, they place some in their holes and runs, either in the sides or thatch; or if you place some of the dead rats so sprinkled beneath the barn floor or the calf-pen, where rats are fond of taking up their abode, it will not only drive them away, but prevent others coming while the dead ones remain. Then, again, I have known them to sew a little red jacket on to a live rat, and start him. Others, again, have fastened a little bell round a big fellow's neck, while others have resorted to the barbarous practice of pouring hot pitch on a poor rat, and starting him in the holes. Either of these plans will have the effect of frightening away the others, for the time being; but they soon become used to it, and will return, if their old quarters are the best.

"When rats take up their abode behind the wainscoting, or in the ceilings of houses, some rat-catchers dislodge them, by first finding or making a hole near the place they frequent; then, thrusting in a handful of salt, and pouring on it two or three spoonfuls of the oil of vitriol, they fasten up the hole. These articles together make such a disagreeable fumigation or smoke, that, to prevent been suffocated, the rats are glad to beat a retreat. Then, again, to drive them from other parts, take a pint of common tar, half an ounce of pearlashes, one ounce of the oil of vitriol, and a good handful of common salt; mix them all together in an old deep pan; then, after spreading the mixture thickly on pieces of brown paper, thrust them into the rat-holes, and let the bricklayer make them secure with broken glass and mortar. The mixture will not only have the effect of driving the vermin away, but keeping them away till the mortar is firmly set. Others, again, I have known to take either smallage seed, nigella, origanum, lupins, or green tamarinds; and by burning them in a room, they say, the fumes have had the effect of driving away both rats and mice. But though all these plans may afford present relief, still they do nothing towards remedying the evil. Then, as to the old methods of destroying them:—Some would take a very porous cork, and cut it into thin slices, or cut pieces of sponge to the size of large peas, and fry them in beef or mutton suet. These, they said, would destroy both rats and mice. Others asserted that unslacked lime, or plaster of Paris, mixed with flour or meal, would effect the same purpose; while some again maintained that steel-filings, made into a stiff paste with meal and honey, would most certainly kill them. Now, whether any or all of these ingredients will have the desired effect, I cannot say, because I never tried them. But I recollect about nine different poisonous compounds, to one or other of which, I believe, the whole profession resorted."

DIFFERENT KINDS OF POISON.

First Poison.—Mix up some flour with honey and a little ox-vomit, till it comes to a paste, then cut it into pieces about the size of dice, and put them into the holes; and it will most assuredly destroy the inmates.

Second Poison.—Take some of the best oatmeal and coloquintida powder, and make them into a paste with honey; then cut them into dice, and lay them in the rat-haunts.

Third Poison.—Mix the seeds of wild cucumbers and black hellebore with such food as they eat, and it will kill them. Or make a stiff paste with oatmeal, honey, and powdered hellebore. This alone will suffice. But let those persons who mix these things be cautious in the use of them, for they are as destructive to human life as they are to meaner animals.

Fourth Poison.—Mix honey, metheglin, bitter almonds, and white hellebore, with flour or oatmeal; make the whole into a stiff paste; then throw pieces into their holes, and it kills them. Some destroy them by putting hemlock seeds into their holes.

Fifth Poison.—Make a paste of bitter almonds, coloquintida, wheaten flour, and honey; then put it into their holes, or lay it where they frequent, and it will certainly destroy them.

Sixth Poison.—Put into their holes hemlock seeds, and powdered hellebore, mixed with good oat or barley meal, and they will so eat of it as to kill themselves.

Seventh Poison.—Take black hellebore, bitter almonds, wild cucumber, and henbane seeds in equal portions; then pound them to a powder, and make a paste of them with hog's-lard and the best oatmeal. This will destroy them both in fields or houses.

Now we come to that most deadly substance, and which is more resorted to than all the rest put together, namely, arsenic.

Arsenical Paste.—Take one ounce of finely-powdered arsenic, one ounce of fresh butter, and make them into a paste with oatmeal and honey; then spread it about those parts of the house they mostly frequent. They will eat of it greedily; then drink till they burst. But as this is a most deadly thing, you should be very cautious in the use of it, and always wash your hands afterwards.

Arsenic Pills.—Take two ounces of fine flour; two ounces of lump sugar, beat to a powder; ten drops of honey; one ounce of arsenic, ground very fine between two marble stones; six drops of the oil of rhodium; eight drops of the oil of caraways; mix them all well together, and make them into a stiff paste, with two or three spoonfuls of milk; then cut it into pills about the size of peas, and lay them where the vermin frequent. But always take up what is left in the morning, to prevent accidents.

Such is the information derived from the long experience of the celebrated rat-catcher, "Whistling Joe." And now, in taking a philosophical view of rat-catching, there appear to me but two principal points involved: first, to get the rats out of their holes; secondly, when out, to keep them out. The first point needs but a little time and patience, and the rats themselves will relieve you of all further trouble; while the second object will be most effectually achieved by Uncle James's imperial rat-trap.

I shall here mention an African mode of poisoning rats, through the means of squill, as stated by a traveller in those parts. When he was residing in Algeria, he heard of a man in great repute among the natives, who sold balls, from the eating of which the rats (which were very large and numerous) died instantly. He was induced to analyze some of the balls of the African alchymist, and found them to contain squill, well dried and finely powdered, with a fatty body, namely, strong-smelling cheese. The gentleman immediately imitated the compound, and says that in more than one hundred trials he found that the rats were killed instantaneously. The prescription ran as follows: "Powdered *Scilla Maritima,* or squill bulb powder, two ounces; and eight ounces of strong-smelling Italian cheese; mix them well together, and serve them out in boluses."

It must be admitted, that to kill rats on the spot by such simple means would be a thing most desirable. But unfortunately, whatever may be the effect of squill bulb powder in Africa, with us in England it is ineffectual; for I have tried it in many cases, and every one proved a failure, at least so far as killing them on the spot. I mixed it in various proportions up to half squill, half cheese; and though every bolus I made, speedily disappeared, still, whether they ate them where I placed them, or carried them to their holes, I know not. But certain it is, our noses soon set us to work; for on taking up the hearthstone, there lay the dead rats.

The following prescript appears in the "Family Economist." "Mix, in equal quantities, the oil of amber and ox-gall; then add to them oatmeal flour sufficient to form a paste, which divide into little balls about the size of ordinary marbles, and lay them in the middle of the room which rats are supposed or known to visit; surround the balls with a number of vessels filled with water. The smell of the oil will be sure to attract the rats. They will greedily devour the balls, and, becoming thirsty, they will drink till they die on the spot."

In the "Agricultural Gazette," a writer states, that having at different times tried many nostrums for the destruction of rats, with very indifferent success, owing to the sagacity of the animals, he adopted a plan which may not be generally known; its simplicity and (as he has always found) its infallibility being its principal recommendations. "Take a red herring, and having opened it, rub the inside with arsenic; then sew it up again, taking care there is none of the poison on the outside; place it in the barn or other place infested, and you will find it quickly disappear, as also the rats. Perhaps, if you were to treat them first with a few red herrings unpoisoned, it might give them greater confidence; and be sure to have dishes and low pans laden with water for them to drink."

The author of the "Cyclopædia of Agriculture," says that in destroying rats in quantities where dogs cannot avail, poison is mostly resorted to; for which purpose arsenic is most generally used. The best method of applying it is, to place slices of bread and butter, sprinkled over with lump sugar, to their holes, or near their haunts, and to repeat this from time to time, until it is readily devoured. The next time, sprinkle it with arsenic as well as sugar. But mind that no other animal can get at it. In this way numbers of rats will be destroyed.

The same writer again says, "Take a quartern of good flour, or the same quantity of good malt, and mix with it an ounce and a half or two ounces of finely-ground arsenic; then add the following mixture of essential oils, namely, ten drops of the oil of caraway, two drops of the oil of anise seed, and one drop of the oil of lavender. But first rub these oils well up in a handful of flour or malt, and then well stir them in with the whole, so that the oils and arsenic may be thoroughly mixed."

Mr. Waterton informs us, that during his travels in South America, an army of rats laid siege to his establishment, and held entire possession of it for years. Not feeling satisfied with prowling at will through the domains, they also rendered the mansion a complete colander, and committed any outrage they pleased, with impunity. However, upon the return of the owner to his castle, the marauders were soon expelled or annihilated; and now, he says, if you would give ten guineas for a rat, you could not find one about his house. The implements of destruction were, oatmeal, sugar, and arsenic. To a washhand-basin full of the best oatmeal were added two pounds of moist sugar and a good dessert-spoonful of arsenic; then, after well mixing them, a table-spoonful or two was laid in every hole: thus the enemy was eventually destroyed.

Here I must take the opportunity of warning you of the dangerous results often attending the use of arsenic in the poisoning of rats. These animals, being poisoned, often crave for drink; and if they should get to any milk, beer, or water, they will sometimes so gorge themselves, as to vomit into the fluid, and thereby produce most serious consequences. A farmer in the county of Limerick used to make butter on a large scale for the London market. His farm was much infested with rats; and, what with the loss of butter out of the firkins, and skimming the milk laid out for cream, he found himself a most extensive loser; so he resolved upon poisoning them with arsenic. And what was the result? why, that all his pigs and calves were poisoned also! It turned out, that when the rats had partaken of the poison they became so thirsty that they went to the milk as usual, and there drank till they were sick; then vomited into the dishes; and the arsenic, being naturally heavy, sank to the bottom. The cream was made into butter, and the skim-milk given to the calves and pigs; and thus the whole were poisoned.

I must also warn you against the use of tallow as an ointment. Some short time ago there appeared in the papers an account of a man whose death arose from putting poisoned tallow to a wound. We have also instances on record of rats being poisoned by candles. A watch-maker was once placed in a most unpleasant position through a rat. His wife had brought into the kitchen a pound of candles, and placed them on the table where she had been sitting. But being called up stairs to answer some one, she was delayed in conversation. When her husband came home they went down stairs; and as the candle was burning in the socket he proposed to light another. She told him there was a pound on the table; but, on looking, no candles were to be found. She was certain she had brought down a pound only a few minutes before. He thought she must be mistaken; but she insisted on the fact. At length an old charwoman who used to clean and puddle about the place was summoned, and questioned as to the candles. She felt indignant, and stoutly denied all knowledge of them. The result was, that the poor old woman was dismissed from their service. However, in about ten days, the place began to smell most unpleasantly. Then it got from bad to worse, till the place fairly stank again. At last he insisted that there must be some dead animal, or something of the sort, under the flooring. The result was, he pulled up some of the boards, and there lay the wicks of all the candles; and under the hearthstone they found a great putrid rat. This of course proved the poor old woman's innocence, and she was reinstated in their service.

But how was it that the rat died through eating candles? The truth was, it was poisoned. It appears it is the custom now-a-days for tallow-chandlers to mix arsenic with the tallow; but for what purpose I cannot say, unless to make them white.

The following recipe is recommended for the destruction of rats, because it is tasteless, odourless, and impalpable. "Take two ounces of carbonate of barytes, and mix it with one pound of dripping." It produces great thirst, and death after drinking; thereby preventing the animals going back to their holes. To prevent accidents to dogs, cats, and poultry, spread it in an iron or tin vessel, hung bottom upwards with wire over a beam, leaving just room enough for rats to pass under easily.

A gentleman living at Hammersmith (as we have before mentioned) lately lost 350 of his favourite song-birds from rats in one night. He tried several means for their destruction, but without effect. At last he hit upon the following plan, which proved eminently successful. He procured some fresh herrings and sprats; then cutting two pieces of stick to use as a knife and fork, he opened the bellies of the fish, and, after lining their insides with finely-powdered carbonate of barytes, laid them down, as if by accident, to avoid suspicion. The plan answered so well that it destroyed the whole rat colony. He says it must be powdered barytes, and the commoner the better, for the purified article is ineffective, as well as costly; and if you cannot get fresh herrings or sprats, a soft Yarmouth bloater will do as well.

A well-known naturalist's prescription for poisoning rats and mice is precipitated carbonate of barytes; not the common carbonate, because that, he says, is as gritty as arsenic, and creates suspicion, while the precipitated is as fine and soft as fine flour. He recommends its being prepared by a druggist, who is most accustomed to such things. But the following is the method, and any careful person can do it:—Dissolve an ounce and a half of common washing soda in half a pint of water; then put into a bottle one ounce of carbonate of barytes with another half-pint of water; that done, pour the soda into the bottle, a wine-glass at a time, and well shake the bottle after each addition. The liquid will become white and thick, and must be thoroughly shaken when all the soda is in; then set it aside to settle quite clear, which may take twelve hours. It will settle much quicker if made with hot water; but then the powder is not so fine, and does not answer so well. When settled quite clear, pour off the water as close as possible from the sediment; then fill the bottle again with water, and shake it well up, to wash out all the soda, &c. from the barytes; let it settle a second time, and when clear, pour off the water as close as possible; then turn out the powder on to blotting-paper to drain and dry. It will clot a little in drying, but is easily rubbed down fine again.

For use :—Take a quarter of an ounce of the powder, two ounces of lard, and a quarter of an ounce of strong cheese; then mix them well together, and spread the preparation on bread like butter. Or, make it up, with flour or meal, into little balls like marbles; or, what is still better, smear it over the rinds of fried bacon, and place it in their holes or tracks. He says, the addition of two drops of the oil of anise seed to the two ounces of lard seems to be more attractive; but he has not observed this addition to do much for mice.

CHAPTER XIII.

PHOSPHORIC POISONS.

THE following recipe for the destruction of rats was communicated to the Council of the Agricultural Society, and is highly recommended as the best known means of getting rid of these most obnoxious and destructive vermin. It has been tried by numerous intelligent persons, and is pronounced to be perfectly effectual.

"Melt hog's lard in a bottle plunged in water heated to about 150 degrees of Fahrenheit; introduce into it half an ounce of phosphorus for every pound of lard; then add a pint of proof-spirit or whisky; cork the bottle firmly after its contents have been heated to 150 degrees; take it at the same time out of the water, and agitate smartly, till the phosphorus becomes uniformly diffused, forming a milky-looking liquid. The liquid, being cooled, will afford a white compound of phosphorus and lard, from which the spirit spontaneously separates, and may be poured off to be used again; for none of it enters into the combination, but merely serves to comminute the phosphorus, and diffuse it in very fine particles through the lard.

"This compound, on being warmed very gently, may be poured out into a mixture of wheat flour and sugar, incorporated therewith, and then flavoured with the oil of rhodium or not, at pleasure. The flavour may be varied with the oil of anise seed. This dough, being made into pellets, is to be laid in the rat-holes. By its luminousness in the dark, it attracts their notice, and being agreeable to their palates and noses, it is readily eaten, and proves certainly fatal. They soon are seen issuing from their lurking-places, in quest of water to quench their burning thirst, and they commonly die near the water. They continue to eat it as long as it is offered to them, without being deterred by the fate of their fellows, as is known to be the case with arsenical doses."

A gentleman who was a great sufferer from rats gives the following as the best method of preparing phosphorus :—"Procure of lard or dripping ¼lb., of phosphorus one drachm, of spirits of wine one gill; place the whole of these in a thoroughly clean pint wine-bottle; that done, then place the bottle in water nearly to the neck. This can be easily done by putting it in a saucepan. Then let it gradually warm, till the dripping or lard is fairly melted; then take it out, cork it firmly down, and shake it, until the contents are thoroughly incorporated. When cold, pour off the spirits of wine, and then it is ready for use."

To prepare it for use, first get some flour, as much as you may think will be necessary, and put in some sugar; then warm the bottle and pour out as much as will make it into thickish dough; that done, cut it in two, and put to one a couple of drops of the oil of rhodium, and to the other, two drops of the oil of anise seed; work them up separately, until the oil is thoroughly mixed; then cut them in little pieces about the size of marbles, and place them in the holes or run of the rats, and there is an end of them.

For the preparation of *phosphorus paste*, Dr. H——'s prescription is as follows :—"Take twelve ounces of starch, and mix it up with half a pint of cold water; when ready, pour in five half-pints of boiling water; it must boil while you pour it in, and the starch stirred at the same time. The instant it forms a jelly, put in one ounce and a half of phosphorus, and cover the basin for a few minutes; then mix it well up together with a wooden spoon, or spatula, till it is cold; after that, add a little powdered valerian root, or a dozen drops of anise seed. Then, to preserve it, put into wide-mouthed jars, and fasten it down air-tight with oil-skin or bladder."

This paste should be spread on slices of bread, and placed near the holes where the rats pass; taking care that they are constantly renewed as they are consumed.

Here let me draw your attention to the imprudent use of poisons, which may be as fatal to human beings as they are to meaner animals. Indeed, where is your own personal safety, if you yourselves purchase the means that, in the hands of a domestic enemy, might prove not only your destruction, by poisoning your food or draught, but be the means of branding you with suicide, on the simple evidence which the murderer himself might adduce, that you were the actual purchaser of the fatal drug. In this refined age of scientific murdering, pray who is safe that has a purse in his pocket or a pound to bequeath? Then, again, as to accidents, it was only very lately that the papers reported a case wherein a girl had been killed through foolishly swallowing two or three pills that had been purchased of some neighbouring vendor for the poisoning of mice; and upon a post-mortem examination the pills were found to contain strychnine. Is it not alarming that quack doctors, peddling apothecaries, and patent medicine vendors should ransack pharmacopœias for the most deadly drugs, merely to poison a poor little mouse? In a word, can the law of the land be said to be a security for human life, while this state of things is allowed? Then, is it not time that the voice of the country were raised, not only against the unnecessary use of poisons, but to petition the legislature to pass laws, placing not only a most rigid limitation on the mere sale of them, but imposing heavy penalties also upon persons for the bare possession of them without a government license? What do we want with such things, either for rats, mice, beer, teas, wines, or candles? But no matter the motive; the excuse ever has been, that they were required for the destruction of vermin. That excuse should no longer be held valid, since phosphoric compounds will answer every end for which such things could be applied, so far as the destruction of vermin is concerned: and as to human beings, it could never be placed either in their food or drink without instant detection, unless they can swallow lucifer matches without knowing it, for its stench will always betray its presence, however small the quantity ; but a dose sufficient to kill a human being would stink the place out ; and I must tell you, that when it ceases to stink, it is comparatively harmless, not having power, I believe, to kill even a rat, much less a man.

The truth appears to be, that phosphorus does not poison at all ; but being conveyed into the warm stomach, it begins to smoke and burn, and thereby creates such an ungovernable thirst, that the animals (as some say) drink till they burst. But I certainly never saw one burst, except by a cart-wheel. It appears to me that the phosphorus creates such an insatiable craving, that the more they drink, the more they want. At last they become so full, that the heart has neither room nor power to act, and so the creatures die.

For the making of *phosphorus compounds*, the following hints may be found useful.

First.—Take good warning, and never touch raw phosphorus with your fingers! If you do, you may take my word for it, you will never touch it after ; for the burn is dreadful, and there is no getting rid of it till it has burnt itself out.

Second.—Whatever quantity of phosphorus you may want, let the druggist weigh it for you. He will cork it tight up in a small bottle with water.

Third.—Get a clean bottle, large enough for the quantity of compound you intend to make. Then melt the lard, and pour it into the bottle. That done, pour off as much water from the phosphorus as you can safely, and turn the phosphorus carefully into the lard ; then pour in the spirits of wine. By this means, you see, you avoid all danger of burning, because you turn the phosphorus out of one bottle into the other.

Fourth.—Keep out the cork, and plunge the bottle up to the neck in a saucepan of cold water ; then place it on the fire, and never leave it ; put your finger, not into the bottle, but into the water, from time to time, to try the heat ; and the instant it is too hot to bear with comfort, take out the bottle, and you will see the phosphorus melted at the bottom; then, after corking it down, give it a smart shake up for about half a minute, and put it by, to get thoroughly cold, so that the spirit may float at the top; after that, uncork the bottle, and pour off the spirit; then the compound is ready for use. But always keep it closely corked, for the more air-tight it is, the better it keeps.

Some recommend keeping the spirit for another time. This may be all very well for rat-catchers, or those who make up phosphoric pills for sale; but for private individuals, who make a little perhaps once a year, or not that, the best plan is to throw it away, and then no harm can come of it.

Mix the compound with flour or meal into pills, as described a few pages back, and you will find them answer every purpose, either for rats or mice. The author of the "Book of the Farm" says, "the phosphoric pills are the best remedy he has met with. It is only necessary to drop a few of them into the holes, and a little trouble of this kind, taken in the heat of summer, when the holes are most easily to be seen, will soon diminish the number of the vermin, if not entirely extirpate them."

Though phosphoric pills are luminous in the dark, and look as if on fire, still they are perfectly harmless. To satisfy myself on this point, I have wrapped them up in highly dried cotton-wool for twenty-four hours, and they did not in the least discolour it. Then, again, I buried some in the best gunpowder for twenty-four hours; and I believe they might lie for twenty-four years, and never affect it. This you can try for your own curiosity and satisfaction.

For the convenience of those who may not wish to make a large quantity of phosphoric compound, I will give different proportions of each ingredient, and then they can choose according to the number of vermin. I must tell you that the first scale will be sufficient for a house, unless the rats are very numerous. But in every case mix up the pills as our Hammersmith friend has described.

First scale.—Two ounces of lard or dripping, half a drachm of phosphorus, half a gill of spirits of wine or whisky.

Second scale.—Quarter of a pound of lard or dripping, one drachm of phosphorus, one gill of spirits of wine or whisky.

Third scale.—Half a pound of lard or dripping, quarter of an ounce of phosphorus, half a pint of spirits of wine or whisky.

Fourth scale.—Three quarters of a pound of lard or dripping, three drachms of phosphorus, three gills of spirits of wine or whisky.

Fifth scale.—One pound of lard or dripping, half an ounce of phosphorus, one pint of spirits of wine or whisky.

For making phosphorus paste, I shall in each scale give the exact quantity of water necessary; therefore, before you begin, measure out the water; and from that, put only the sixth part to the starch, to soften it; then put the other five parts on the fire to boil. In the meantime mix up the starch. But be sure to put them into a round-bottomed vessel. A flat-bottomed one will not do, because the phosphorus will get round the edges; and by its hardening there it will render the paste useless. Now, when the water boils, pour it into the starch, stirring all the while; and the instant it becomes a jelly, put in the phosphorus, and cover up the vessel for a minute or two, to let the phosphorus melt; next beat it up well till it is cold, and then it is ready for use; but to preserve it you must make it air-tight. Spread it on thin bread, like butter, and do not forget the scenting oils.

First scale.—One ounce of starch, one gill of water, one drachm of phosphorus.

Second scale.—Two ounces of starch, half a pint of water, quarter of an ounce of phosphorus.

Third scale.—Quarter of a pound of starch, one pint of water, half an ounce of phosphorus.

Fourth scale.—Half a pound of starch, two pints of water, one ounce of phosphorus.

Fifth scale.—Twelve ounces of starch, three pints of water, one ounce and a half of phosphorus.

Sixth scale.—One pound of starch, four pints of water, two ounces of phosphorus.

I shall here notice the best kind of bottles for phosphoric compounds, and also give some general cautions for using them. In the first place, never introduce more phosphorus than the proper proportion; for if you do, your object will most certainly be defeated, because the vermin will not touch it. I have found small pickle-bottles the best, both for mixing lard and phosphorus, and for keeping the paste in. They are wide at the mouth, and thereby do away with the necessity for melting before the fire, because you can scoop out the mixture with a knife, spoon, or flat stick. Still, always remember, not only to bring it tight down, but to tie the bladder over it, and keep it in the dark. In this way it will keep good to the last. I must also tell you, that in the absence of spirits of wine or whisky the same quantity of good gin, rum, or brandy will do.

Phosphorus and lard may be mixed without spirit. But in that case you must take the proportions of either scale, and serve them just the same as you would with the spirit, with one only exception, that is, that when the ingredients are fairly melted, you must cork down the bottle, and shake it till the lard begins to set. But the distribution of the phosphorus is not so complete as it is with the spirit, because some of it will be found in small beads at the bottom, whereas the spirit has the effect of entirely diffusing it, in the most minute particles, through every part of the dripping or lard.

If at any time you wish to divide the phosphorus into equal parts, turn it out into a saucer of cold water; then hold it with a fork, and cut it gently under water with a knife. When this is done, and the phosphorus put away, be sure to stick the end of the knife and fork into the fire, to burn off the particles; and then no harm can come of it. At all times keep the phosphorus under water; and, whatever you do, keep all fire from it, for it will catch instantly if out of water, and burn most furiously! Consequently the safest plan is only to buy what you want for present use, and then all danger is obviated.

To prevent accidents to the live stock, always place the poison sufficiently down the holes; but in barns it is a good plan to have a large rough box about six feet long, fifteen inches deep, and a yard wide. At the lower corner of each end, near the wall, cut a hole about three inches wide, for the rats to pass in and out easily. When the box is well placed, raise the lid, and strew the inside with corn chaff; then put in low vessels filled with clean water; or, if you like, you may give them a treat or two of skim-milk. But, to draw them in the first place, trail a red herring round the floor, close to the wall, through each hole into the box, and there leave it. They will eat it, and then drink of the water. Look to the vessels every morning; and you may throw down some bread crumbs, scraps of cheese or bacon-rinds; but be sure each time to cover the box loosely over with straw, taking care to keep the holes clear; and when you find them eat and drink freely, then lay down your poison, spread roughly on a wooden platter, and not in pills or boluses. By this plan they cannot carry it away, but must eat it where it is, and will afterwards drink till they die. Bury the dead every morning in the dung-heap, so that pigs or other animals cannot get at them. Replenish the poison and water, and by this means you will soon clear the place.

And now, in summing up the matter on poisons, let me tell you that until the law, or universal consent, shall put an end to the use of arsenic, barytes, nux-vomica, and such like things for the poisoning of rats, &c., of course you have an undoubted right to use them, if you think proper. But unless you are on good and friendly terms with a druggist, you must have a written recommendation from some qualified medical man who can guarantee your intentions. Without this, no druggist can sell you such things without being subject to a heavy penalty.

CHAPTER XIV.

MISCELLANEOUS ANECDOTES.

A Barrack for Rats.—An extensive bacon merchant, who kills between forty and fifty thousand pigs in a season, has adopted the following successful method for destroying the rats which abound in his premises. He has erected a square building, eleven feet long and seven feet wide, with a wall three feet high, having stone flags laid flat on the top, but projecting over the inside of the wall. All round the wall, on the inside, at the base, are numerous small holes, like pigeon-holes, which do not go quite through, except a few, to allow free passage for the animals. Outside of the barrack is a plentiful supply of water and food, such as bones and useless offal. The interior of the walls is occupied by boards, lumber, and straw,—just such concealment as these animals are known to like,—and the whole is covered with a movable wooden roof. When it is judged proper to destroy them, the passages are stopped at the outside, the roof is lifted off, and the boards are taken out. The frightened animals run up the wall, when they strike against the projecting flags, and fall back again. They then run into the small holes below; but these are only just large enough to admit their bodies, whilst their tails remain sticking out, a secure prize for the men, who go in over the wall; and by this unlucky appendage they suddenly drag them out, and fling them to a posse of anxious dogs outside of the fortress, or into a tub of water, where they are soon destroyed. As there are not holes enough in the wall inside, the noise and uproar soon frighten another division of rats into the vacated openings, and these being treated in the same unceremonious manner, the whole garrison is thus speedily destroyed. As many as seven and eight hundred have been killed in a single clearing.

A Fox in the Loft.—Rats being fond of straw, they become very numerous in the lofts where this article is kept for singeing bacon, and they cut it into short pieces with their teeth, which renders it useless for this purpose. A bacon merchant tried the effect of putting a pet fox into the loft to mount guard, and it was found that he killed such numbers of the rats, that four more were procured to garrison the place, instead of one.

A Fox in the Barn.—A Scotch gentleman was very much pestered with a horde of rats about his premises. The barn and dairy suffered most alarmingly from their depredations. Lately, however, he got a young fox, and shut it up in the barn for a few days; when the rats, disliking the stranger's company, removed in a body, and pitched their camp in the dairy. Of course, the gentleman was equally anxious to start them from among his butter and cheese; so he chained master Reynard at the end of the building, when the rats took huff, and emigrated from the place altogether.

The "Farmer's Magazine" gives the following recipe for keeping vermin out of ricks:—

"Take a pound of nitre, and a pound of alum, and dissolve them together in half a gallon of spring water; then get a bushel of bran, and put in one quart of the liquid, and mix them well together. When you build your ricks, every second course, take two handfuls of the mash, and throw it upon them till they come to the easing." An agricultural gentleman states that he has found it so effectual, that he never has a rick put up in any other manner.

How to destroy Rats by Suffocation.—Rats may be destroyed in great numbers in barns, in the following manner. Before the corn is removed from the ground, get some common iron chafing-dishes, and having placed each on a couple of bricks, fill them with lighted charcoal; then strew on a quantity of broken stick brimstone: this do as quickly as possible, and quit the barn while holding your breath; if not, you may be suffocated also; close fast the door, and leave the building shut for the next two days. On re-entering the barn you will find numbers of rats lying dead round the chafing-dishes. Some may die in their holes, and will cause a most offensive nuisance, if precautions be not taken to prevent it. This is easily done by stopping up all the holes with mortar. Perform this operation again the following harvest, just before storing, and you will no longer have reason to complain of annoyances from rats.

To keep Rats out of Barns.—A bed of shingle, or very coarse gravel, from six to ten inches deep, beneath the floors of aviaries, barns, or any places infested with rats, will effectually keep them out; for they cannot, or will not, burrow in shingle.

To keep Rats out of Warehouses, &c.—The proprietors of the bonded corn-warehouses on the banks of the Thames, to keep out the rats, have the wooden flooring, and the under-parts of the doors, lined with sheet-iron; and the foundations of the buildings are sometimes set in concrete mixed with broken glass. These are materials too hard for even the teeth of rats to deal with, and so the corn is preserved.

Rat-catching in the Gambia.—The author of "Elements of Natural History," tells us, that in warm climates rats grow to an enormous size, and that she has been, not only a witness to their depredations, but a sufferer by them during her residence in tropical climates, and it was with difficulty that she kept them from her sleeping apartments. A vessel in which she sailed could only be cleared of them by sinking it under water for a time. A rather novel way of hunting them was practised in the Gambia, where they used to make numerous nests in the fire-places and chimneys, which were erected against the damp, and not against the cold. This lady says, that she was one day seated at work in her bed-room, and, hearing a slight noise, turned her head, and there was a great rat on the table close to her, apparently watching all her movements. She continued her employment for a few moments, so as not to alarm the gentleman; then stole gently away, and shut the door after her. She called the servants, and in a few minutes a regular hunt was commenced. Two of the party, armed with sticks, turned the rats out from their hiding-places; and, as they invariably ran round the sides of the room, other servants were stationed at the corners, who squatted down with towels in their hands which they held open for the rats to rush into. They did, this in every instance, when their necks were immediately twisted, and they were thrown aside to be ready for another victim. In this manner, and in this one hunt, they succeeded in capturing and destroying more than half a hundred rats.

A FEW CONCLUDING HINTS.

And now having, to the best of my powers, given you a most minute description of the nature, fecundity, and devastating powers of the rat, with the best means for its extirpation, allow me to conclude with a few words of friendly advice.

In the first place, let us have no more grumblings and poverty-striken yarns about bad harvests, high rents, low prices, and all such humdrum; but let farmers buckle up their belts, and set to work with the enemy like Englishmen, in downright earnest, and then will they be able to take the full produce of their crops to market, and bring back the guineas in their pockets. To me it matters little as to whether farmers become their own rat-catchers, or employ a professional man at a reasonable salary. But I do wish to impress it most emphatically on the minds of their sons, that the rat is the greatest enemy they have. I say their greatest enemy, because in too many cases it blights their hopes and prosperity by gradually devouring their fathers' substance, and thereby keeping them in continual needy circumstances, and, in too many cases, entirely beggaring him. Therefore, I say, let the sons take example, and never do another day's work, or see their sweethearts, till they have destroyed every rat upon the farm. I have laid down the plans, plain and simple,—it now wants but the will; and if their sweethearts be sensible girls, they will never grant them another favour or kiss till they have done their work of destruction right manfully, when, perhaps, they will pay them in full, both for yesterday and to-morrow.

And now have the kindness to give Uncle James's best respects to your amiable wives and daughters, and tell them never to give you an hour's rest, night or day, till you have bestirred yourselves, and put an end to their privations, and, in too many cases, comparative poverty. Why should they be deprived of the little comforts of life, merely to feed a colony of hungry, destructive vermin? Nay, more; pray, what right have you to keep these animals, not only to destroy your own properties, but for the destruction of your wives' and daughters' Poultry also? Put an end to this state of things, and then will your wives and daughters be able also to go with merry hearts to market, and buy of everything, even more than they want, out of the profits of their own poultry-yard.

PART III.

SOME PROFITABLE HINTS ON THE BREEDING, FEEDING, AND GENERAL MANAGEMENT OF POULTRY.

HAVING so far disposed of the destructive vermin of the farm, allow me to give the ladies a few friendly hints on the Management of the Poultry Yard; and the first question to be asked is, Which is the most profitable to farmers, the feeding of cattle or of poultry?

Some time ago, a poultry breeder inserted an article in the public papers, stating that it takes five years to raise an ox for the market, and that he will undertake to feed the same weight of poultry for the market in three months, and at half the expense. That is to say, if the ox weighs half a ton, he will feed double that weight, namely a ton of poultry for the table, in three months, for the same expense as the ox has cost for victuals during the five years. The consequence would be, were they both to start feeding together, that at the end of five years, where the farmer would have made one pound profit, the poultry feeder would have made forty; or, where the farmer would have cleared five guineas, the poultry feeder would have cleared two hundred. Here it is clearly shown how much more profitable it is to breed poultry than cattle, if proper pains be taken.

I am sorry to say that, in the great majority of cases, farmers' wives and daughters do not manage the poultry yard properly; that is to say, they do not manage the breeding as they ought to do. For instance, they seldom or never think of introducing new blood among them; but let fathers, mothers, sisters, and brothers, grandchildren too, all live and breed together as they like; and as the old cocks die away, their sons take their place, and so they go on breeding in, one generation after the other. The consequence is, they dwindle down in size and quantity, till at last their eggs are few and very small, and the hens themselves no bigger than chickens ought to be. Pray, are prize cattle, pigs, and sheep produced in this way?

The Bantam Cock.

Some years since, at a roadside inn, in the neighbourhood of Oxford, I had an opportunity of witnessing the effect of this in and in breeding. The landlord had a handsome lot of fowls of the spangle Polish breed. Their size and beauty attracted my attention and admiration. I was in the habit of seeing them daily. Presently, however, some one made him a present of a bantam cock. He put it down in the yard with the other fowls, when a fight instantly took place, in which the bantam, after a hard struggle, proved the conqueror. The landlord was so pleased with the little fellow's courage, that he kept him, and killed the other. The result was, that in the space of four years, the whole stock was reduced to a race of mongrel-bred bantams, and very little bigger than partridges. Now there were two reasons for this. In the first place, no other cock was kept; consequently the bantam was breeding with his own offspring down to the third generation. In the second place, when a fowl was wanted for the table, they always took the largest, and left the smallest for stock.

Farmers' wives should do the very reverse of this; they should always keep the finest birds for stock, and introduce two or three of the finest cock birds that friendship or money can purchase, keeping all the largest eggs for sitting; by this means they will have a yard of poultry that will pay them handsomely for their trouble. Moreover, if farmers' wives in general would do the same thing, then could they exchange birds with each other, and the finer and more distant the male birds are from your own stock, the finer and more vigorous will be the chickens.

Qualities necessary to make a good Fowl for the Farm.

I must tell you that, for my own satisfaction and the benefit of farmers and their families, I have tried various experiments in the cross-breeding of fowls, and it may be some satisfaction to know the results of my experience.

In the first place, let us clearly understand what are the qualities necessary to constitute a good fowl for the farm, because, in this case, they are not wanted for fancy, but for profit. Well then, to be profitable, they should be good foragers, good layers, good sitters, good mothers, and good large chickens for the table. These are the qualifications necessary to constitute a good profitable fowl for farmers' wives and daughters.

Fancy Fowls.

In the course of my experience I found none of the fancy fowls (either Hamburgs, Polish, Malay, Spanish, or Game) possessing the number of points necessary for perfection ; and as for Cochin China, whether half-bred or whole bred, whatever you do, avoid them, for they are dear at a gift. They may be just the thing for the Slave States of America, where each fowl can have a little black boy to wait upon it ; for unless the food be brought to the bird, or the bird be carried to the food, it will do without it, and will scarce deign to move, even when pinched by the fangs of hunger. It is a perfect drone in all its actions ; and when it crows, it is just about as musical as a donkey braying through a speaking-trumpet, or a cow-boy calling up the cattle with a cow's horn. They appear huge birds ; but when you lay hold of them they feel as if they were made of cork, and you have to grope about after the body, being nothing but skin, bone, and feathers. When they are killed and plucked, they are one of the sorriest spectacles imaginable. In this respect the Cochin fowl and the barn owl ought to go together ; for one looks quite as tempting as the other. All that ever came in my way looked as hollow, flat, and skinny as so many half-starved ducks, and had scarcely an ounce of eatable flesh upon them ; and even that was, comparatively speaking, as coarse and tough as Buffalo beef. The result is, that in no shape or form whatever are they a fowl for the farm. They may do very well for gentlemen's lawns, because they never fight or break down the shrubs ; or for citizens of London, who turn a back garret into a poultry-yard, and by this means sometimes get a new-laid egg for breakfast; for it is a well-known fact that some of the finest specimens of these birds have been produced in this way; that is to say, the fowls themselves have been hatched and reared in the garrets of London.

Spanish Dorking Fowls.

You may now perhaps be anxious to know what kind of fowl I should choose for the good ladies of the farm; and I tell you, before all fowls in the universe give me a breed between the Spanish and Dorking! The cockerels make the most noble, handsome, dashing birds that can be desired, and as strong as lions; while the pullets make most handsome, full bodied, sprightly hens, looking more like small turkeys than barn-door fowls. For laying and breeding they are not to be equalled; and for the table they may challenge anything that can be produced. They are both large and plump, delicate as a curd; at the same time short-eating, juicy, sweet, and tender. In a word, I believe them to be the most delicate and delicious fowl, whether as pullet or capon, that can be placed upon a platter.

I have had some of these birds that lay the entire year round, not stopping even to sit or moult; but these of course were extreme cases. The May chickens would all lay at or before Christmas; and some of the June birds have done the same. The hens would each lay from 120 to 150 eggs in the year, regulated of course according to their sitting and moulting; and their eggs would average six to the pound, or, if you picked them, five would weigh a pound weight. But I must tell you, that there was as much credit due to Aunt Jane's management as there was to the birds themselves; for it does not signify one rush what the breed may be. If they are badly managed, there is no pleasure or profit attending them. I have known some silly people who, having a few pent-up fowls, strive to make them profitable by what they called economizing the food; that is to say, half starving them. Now, to say nothing of the cruelty, just let such wiseacres take an egg in hand, and tell us how it is possible for a fowl to produce such a large substance as that, four or five times a week, when, in the same period, it does not receive scarcely half the weight in food. Such persons may be wonderfully clever in building castles in the air; but they cannot build either a house or pigsty without sufficient materials. Then some there are who seem to believe that both pigs and poultry thrive best in the dirt. Never was there a greater mistake; for experience has taught me to know, that, to produce profitable or prize pigs and fowls, the points to be most scrupulously observed are, extreme cleanliness, proper food, and regular feeding, perfectly dry situations, pure air, and plenty of pure water. With these observances anything will thrive and do well; but the instant you neglect them, that instant they begin to go wrong. So the truth is, that the real method of economizing poultry, is to feed them well, but not to waste; and then will they not only repay you plentifully with fine eggs, but supply the table abundantly with fine fat chickens.

Neglect and Mismanagement of Poultry.

Curiosity one day led me to go in quest of a weasel; and after a fruitless hunt I called at a roadside house to take lunch, where I fell in with a neighbouring farmer, who was much taken with my dogs. I asked him if he could inform me where I was likely to find a weasel. "A wissel," he exclaimed, with a volley of oaths. "Why we're swarming wi' 'em; they're eating us up alive! Now, do you know, sir, that with young-uns an' old-uns, they've eaten an' destroyed over 200 head of chicken this season in my place alone! As to eggs, we can't get one, except what we find in the hedge rows; an' now there's not a fowl will go near the house; they roost in the trees, and elsewhere round about. 'Twas only last evening I set my chaps to stone 'em out; but, lor, it's no use; they only fly right away, an' settle in other people's corn. But I'll tell you what it is, sir, ——" Here he heaped all the hard words which none but a drunken, foul-mouthed man could utter, upon the head of a poor little speckled hen, calling her everything but a lady, because she led the others astray. The result was, I went to see his farmstead, and there was nothing but a squalid recklessness in the appearance of everything around. The truth was, the man was a determined drunkard, and sold and spent everything he could lay his hands on.

When I arrived at the hen-house, I certainly never saw such a pestiferous den in my life. The thatch was a black matted mass of rotten straw, half falling, and filled with rat-holes. On the inside there was nothing but the beams for the fowls to roost on, and open to the weather. These beams, of course, formed excellent foothold for the rats, on which to attack the fowls while sleeping. The dung had accumulated into heaps; some of which were at least a yard high. Around them lay the remains of four or five fowls, half-eaten, and the remainder rotten, and one mass of maggots. The place was drilled like a colander with rat-holes. I asked the man to step in and see. "No, hang it," said he; "no, I shall be smothered wi' fleas, mon;" and he spoke the truth, for I was literally covered with them; and altogether I certainly never saw such a desolate heap of contaminating filth in all my life.

His wife and two fine young girls, his daughters, came forward with deep-seated grief stamped upon their countenances. I asked him, in a familiar manner, how he would like to roost there himself. His only reply was a loose drunken shrug of the shoulder. How then could he expect the fowls to like it? I then asked him, why he allowed the dung to remain as it was. "Oh!" said he, "it makes excellent manure, mon, when well rotted." "But why not let it rot outside?" "Uts dart un," said he, "I never give that a thought afore." "And what water have the fowls, pray?" "Oh, they drink out o' th' pond," he replied. This pond was a thick, deadly-looking puddle, that received the drainage from the dung-heap. I then inquired why he did not send for the rat-catcher, and have the place cleared of the vermin. Here his wife spoke, and told me that she did not know what had come to him; but for the last two years he had taken to drinking, and was always at the ale-house! That he had deprived her and her family of the benefits arising out of the poultry; and that he sold every bird he could lay his hands on for mere drops of liquor. As for the rats, she said the place was swarming with them, because he refused to give the rat-catcher more than five shillings a day, instead of ten; and the man would not come. Thus it was, that everything around them was going to rack and ruin. Here the poor daughters buried their faces in their aprons, and turned their backs, while he roared out a volley of oaths against all the rat-catchers in the universe; "an' if he had his will, he'd hang every man of 'em." His poor wife, with tears in her eyes, tried to reason with him kindly, and received nothing but hoarse bullying in return. "But, my dear," said she, "the gentleman says they are rats." "I don't care for you or the gentleman either; I say they're wissles. Who ever heard of rats killing hens and chickens? I suppose you'll say next, that rats eat the eggs! Warn't I born an' bred a farmer; an' think I don't know a rat from a wissle?"

"Come, my friend," said I, "we'll put this matter to the test, and see who is right." I then pulled a ferret out of my pocket; having brought with me the smallest I had, thinking that if I fell in with a weasel, and it should take to ground, perhaps the ferret might unearth it. I bid them all stand back. Then placing my dogs at the most likely holes outside, I went into the roost, and put down the ferret. It was not long before out dashed the rats. The dogs succeeded in killing some dozen or twenty; but there being only two dogs, several of the rats made their escape, while others turned back into the holes. "There, sir," said I, "pray are those weasels or rats?" He was like the majority of ignorant, self-conceited men, who, if the fact be placed before their eyes, will not acknowledge it, but turn the subject. "Ah, well," he said, "it don't much signify, they bring so many French eggs to market now, that poultry warn't worth keeping." Then clenching his fist, "I'll tell you what 'tis, sir, the French can always undersell us!" "And well they may," I replied, "if the British farmers neglect their poultry as you do. And now, sir, let me tell you, that indolent men are always glad to cling to any excuse to cover their laziness. But, on the other hand, sir, I am satisfied, by practical experience, that if the farmers of Great Britain will only allow their good wives and daughters their just rights, namely, the profits of the poultry-yard, they will, by proper attention to breeding and general good management, produce both eggs and poultry that for size, price, and quality may bid defiance to any and every thing that can be imported, either from France or any other country."

"That, sir," said the wife, "is certainly most true; for when he allowed me to manage the poultry I always found plenty of customers for everything I took to the market; and I'm sure there were as many foreign eggs brought over then as there are now. But the truth is, sir, he never pays the least attention to it, except to catch what he can, then take it down to the Rose and Crown, and there sell it for what it will fetch. You see, sir, there is not a turkey, duck, or goose about the place; and as for the few poor fowls, he has not given them a handful of grain for months!" "I knows that," he grunted out, "and why should I? They won't fetch nothin'—so let 'em grub for themselves; and if they won't, let 'em die." With this he turned on his heel, and staggered off to the ale-house.

It is needless to say anything more about this man's pot-house observations; but suffice it to say, that after replacing the ferret in my pocket, and giving the poor wife and daughters some wholesome advice, I went my way, and felt perfectly satisfied, from what I had heard and seen, that that was not the way to manage poultry.

However, I must tell you that I met the good woman about three years afterwards, and she told me she was a widow; that about a week after I was there, her husband, in coming home from the Rose and Crown late at night, tumbled into a ditch, and being stunned in the fall, was drowned in a foot and a half of water. "But, sir," she said, "it is indeed an ill wind that blows no good. Her friends and neighbours rallied round her; and now there was not a happier family in England. Everything prospered with them, and she had the finest and most profitable stock of poultry of any one in the country, and could sell ten times more if she had them. But oh, sir, we shall never be able to repay you for your kindness in showing us how to make phosphorus pills, and the advice you also gave us in poultry-keeping. Uncle James is ever in our mouth, for if any fowl goes wrong, or a rat makes its appearance, the cry directly is, what will Uncle James say, if he comes and sees it? The truth is, sir, we are never without the pills; I make a quantity, and keep them by me in close-covered jars, and as one lot is used I make another; and so we keep the vermin under. My sons and daughters have each a sweetheart, and they make it a pastime, now and then, to go all over the farm, and drop some into every hole they can find; nor do they ever go to work without taking some in a bottle with a spoon; but should they happen to forget, they'll sing out from the other end of the field for Uncle James's pills, for that is the name they have given them. And now, sir, I do not think there is such a thing as a rat or a mouse upon the whole farm; and as for cats, the instant one comes about the place my sons shoot it. The consequence is, sir, that we seldom or never lose a chicken, or, indeed, anything else; and thus are we enabled to live both comfortably and happily."

Aunt Jane's Management of Poultry.

Aunt Jane had a splendid stock of Spanish Dorking fowls; but they were more for quality than quantity. In her management of them she always scrupulously observed four cardinal points, namely, extreme cleanliness; proper food, according to the season, and regular feeding; pure air; and plenty of pure water.

Her hen-house was made of wood, with a boarded flooring and a tile roof. There were openings in the tiles, but so constructed as to keep out the rain. There were four trap-doors, one on each side of the house, four feet square; that is to say, four feet from corner to corner, which could be opened and shut at pleasure by pulley-ropes inside. These she regulated according to the weather; thus could she have a thorough current of air through the place for any time she pleased. In summer she would leave one open all night, because it was beyond the reach of foxes, and in winter she kept them perfectly close and air-tight. On one side of the house she had a shed, in which there was a cart-load of old mortar and rubbish, taken from the pulling down of an old house, and beaten fine. In this the fowls would scratch and roll for the hour together, to work the dust in among their feathers. It is the finest thing known, to cleanse the bodies of birds from all kinds of vermin. In the nest holes she had sliding boxes, for the convenience of changing and lime-washing. The perches were very thick and round, and so arranged that the roosting fowls could not soil each other. She always did the work herself, or saw it done; "and then," to use her own words, "she knew it was done properly," and if done regularly, there was but little to do; but if she trusted it to any one else, then the fowls might do it themselves. She had the house cleaned out every morning; and in hot weather she changed the nests once a week, using but little straw; that is to say, she changed so many each morning, and all the clearings out were taken directly to the dung-heap. She never allowed her brooding hens to sit in the hen-house; but, not having an orchard, she had a shed built in the garden purposely for them, and properly partitioned off, so that the fowls could not see each other; and thereby they were kept from fighting. As soon as a hen brought forth her chickens, the old nest was burnt, the box lime washed, and a new nest put in. By these means she always kept the vermin under, which otherwise, she says, not only drive hens to lay astray, but cause sitting hens to spoil their eggs, by jarring and shaking them, through wriggling and twisting about on the nest, while searching after the tormenting fleas.

Her coops were like large boxes, made of stout inch fir, two feet six inches square, and two feet high. The back and front opened up, so that they could be propped up or thrown back on the top. Both sides were wired in with strong iron wire, set wide apart, with the centre ones loose, so as to slide up and down, to put the hen in, or let her out. These coops I considered an excellent contrivance, because in cold, tempestuous weather, she could shut down one flap; so that it stood with its back to the wind; and these flaps also sheltered them from both sun and rain, while the wooden bottom not only prevented the hen from scratching, but formed a nice dry spot for the chickens to clutch on. Nor was this all; for when evening came, and the hens had gathered their chickens under their wings, Aunt Jane would fasten down the flaps for the night; and then they were safe, not only from the weather, but from the attacks of vermin also.

I must tell you, that on one occasion, some five years ago, she had a narrow escape with all her brood-chickens. She contracted with a rat-catcher, for ten shillings a month, to keep her free from vermin; and, though he did his work most faithfully, still, one night in spring, a body of migratory rats came and attacked her hen-coops. Now, whether the stout inch fir was too tough a job, or whether they were disturbed, she cannot say; but they left their teeth-marks in every one of the coops, and she believed that the reason they did not attack the hen-roost was, because her faithful dog had his kennel close to the hen-house door, for the purpose of keeping away both foxes and thieves of every other denomination. In the morning she sent for the rat-catcher, and when he came he found the rats had taken up their temporary abode in the faggot-stack, where it was believed the dog had driven them, as several were found dead about the place. The men and neighbours were summoned to form an outer ring; and to remove the stack was a matter of short work. Suffice it to say that, with the aid of dogs and sticks, not a rat escaped, though they amounted to hundreds. However, to secure the chickens for the future, she had the outsides of all her coops bound with tin, eight inches deep from the ground.

Aunt Jane always fed her poultry after she had had her breakfast, when most of them had returned from clearing the hedges and grounds of all the worms, snails, slugs, grubs, and insects that came in their way, by which they not only rendered the farm most essential service, but supplied themselves with animal food—a matter of the utmost importance for good laying and general good health. The rest soon came at the well-known sound of her silver whistle, and then she gave them some barley, or the refuse from thrashing; the same again in the afternoon, before going to roost. As for the brood-hens, she always opened up the coops, and supplied the chickens with food and water as soon as she arose. After dinner, she would amuse herself by setting them running, like little race-horses, after pieces of fresh meat, cut very small. Then, again, I have often seen her, with spade in hand, turning up the loose mould for worms, and her swarming family scrambling around her. But, as they became big fellows, and began to scratch, and were too unruly for the garden, they were taken to the hen-house, where they soon began to roost, and take their chance among the rest.

This was her method of feeding poultry during the spring and summer months; but in the autumn, when the slug and insect tribes became scarce, and the fowls were in the moult, they required a more substantial and nourishing food than mere grain, to enable them to go through the moult vigorously, and this could only be done by giving them a little animal food.

In addition to the refuse of the kitchen, Aunt Jane used to contract with a butcher in the neighbouring town, to supply her twice or three times a week with a certain amount of offal from the slaughter-house. It did not signify what it was, so long as it was animal substance, and healthy and sweet; for fowls are perfectly carnivorous, and will eat the flesh of anything, from a horse to a mouse. Not that I recommend horse-flesh, though I have seen people, both in London and the suburbs, feed their fowls with the flesh they had bought of dog's-meat men, out of their barrows. Aunt Jane did not do that, but fed them with livers, lights, clean-washed paunches, fat guts, or indeed any pieces the butcher might send, after slaughtering oxen, sheep, or pigs. The whole was boiled till perfectly tender, then cut up into small pieces, and placed with the liquor in wide dishes; and when sufficiently cool it was laid down for the fowls. This she did every other morning. At other meals she fed them with grain. They devoured the meat, and sucked the warm liquor with the utmost relish; not drinking it as they do water, but kept their beaks in, and sucked as long as they could hold their breath, and then at it again, till they had had their fill. This system she continued through the autumn.

In the depth of winter she always gave them two warm meals a day. She had the same quantity of meat each week; and, instead of giving them a full meal of it every other day, she divided it, and gave them half each day. That is to say, she had half the meat boiled tender, and after taking it out, and cutting it up small, she took one half of the cut meat, and mixed it with equal portions of barley-meal, sharps, and boiled potatoes, sufficient for one meal; then poured on some of the boiling liquor, till it became a thick pudding; and when sufficiently cool she would let them have it for supper. In the morning the other half of the meat was made to boil; then mixed with barley-meal, potatoes, and sharps, as before, and that served them for breakfast. By this method they had only a small portion of animal substance each meal. But when the snow was on the ground, or in hard frosty weather, she had the food taken to them in the hen-house; so that they could please themselves as to when they would come out. By this sort of attention and treatment she was amply rewarded; for the February, March, April, and May, and even some of the June chickens, would lay at and before Christmas. Thus was she supplied plentifully with fine new-laid eggs nearly all the year round; and besides that, she had plenty of fine fat capons for the table.

Should a fowl, at any time, become ill, it was instantly removed, to prevent mischief spreading among the others; still that was a matter of very rare occurrence. But Aunt always maintained that pure air and cleanliness were quite as necessary as good food, for the preservation of health. Besides perfect ventilation, she considered it was also necessary to lime-wash the boxes, and change the nest. But, "to make assurance doubly sure," she, every Monday morning, during the hot weather, had the floor of the hen-house strewed with unslacked lime, powdered fine, and then watered from a watering-pot; and after letting it stand for a time till settled, it was swept clean out, and the place mopped dry. This process annihilated the fleas, and did much to purify the place from anything like disease. As to the water they had to drink, that was always kept in the shade, and supplied fresh every morning in clean earthen dishes.

Now, I have told you Aunt Jane's methods of managing poultry; and though she was most scrupulous in the observance of every rule, still she was equally so with the dairy, and her domestic duties. She superintended everything herself, and was never confused, or in a hurry, because she had a time for everything, and everything was done in its time. This always kept her placid and happy. Nor had Uncle any cause of complaint, since his buttons, shirts, and stockings were ever ready and in good order!

Breeding of Spanish Dorkings.

Having seen both the wrong and the right methods of managing poultry, let us set to work to obtain the finest and most profitable breed we can keep; namely, Spanish Dorkings. Pure-bred Spanish fowls will not serve your purpose, for three reasons: first, they are not good foragers; second, they are the worst of sitters; and third, they are very hard moulters; but for laying, they far outstrip all other fowls, and their flesh is very sweet, delicate, and tender eating. So by crossing them with the Dorking, you overcome all their bad qualities, and retain their good ones, and thereby obtain a fowl possessing a greater number of valuable properties than any other fowl that is bred.

The first thing to be done, is to provide a suitable place for them. If you have not a place where you can turn them out without other fowls getting at them, you must build a small place, and lath it in securely, to prevent all mishaps whatever. Buy your hens in the autumn; then, by keeping them through the winter, after Aunt Jane's manner, they will have full bodies and sound constitutions for laying. But clearly understand me; if the hens you buy are laying, or about to lay, their eggs will be of no use to you that season, because it is a hundred to one but they are all contaminated: so you must take care of them, and pen them up directly after the moult, for the next spring.

In the next place, get six of the largest and finest Dorking hens you possibly can. Those coming in two year-old are the best; but if you cannot get that number to suit you, choose those you like, and make up the number with the finest maiden pullets. Or, if you please, have them all maiden pullets, if they are of superior size and quality; but those coming in two year-old are the best for the coming spring, because they have arrived at maturity, and produce finer chickens. I must also tell you that I have raised some most noble birds from three year-old hens. However, in this case, buy with a liberal hand. If you do not, you will find it penny wise and pound foolish work. I knew a person of this kind, who studied what he called economy in the matter, and was going to carry all before him, with Spanish Dorkings; he gave half-a-crown a-piece for six flimsy-looking little hens, and because they had five toes each, he set it down for certain they must be thorough-bred Dorkings. But, for the purpose he wanted them, they were not worth twopence a dozen. You must first look after the largest, smartest, plump-bodied solid-looking hens or pullets you can find, and then talk about the price. But give what you may, within reason, they will amply repay you by the noble race of chickens they will raise around your farmstead.

With respect to the choice of a Spanish cock, there needs no hurry, since any time before Christmas will do. But I must warn you against those small pigmy Spaniards that many little fanciers possess, and over which they are very conceited, because they have white faces. They will tell you that the twenty-fourth or twenty-fifth great grandfather of these birds cost three, four, or five guineas. All this may be very true, and the grandfather may have been a very fine bird. But what have they done? They have bred them in and in with each other for so many generations, that, comparatively speaking, they have dwindled down to mere cock-sparrows. And as to their having white faces, what is that to you? you are not looking after faces; they may be all white, half white, or all red, for what you care. What you want is, a fine, dashing, large-bodied, thorough bred, two-year-old Spanish cock; you care nothing about the colour either of his face or feather. You want him for size and breed, and not for fancy. Such birds may sometimes be met with in the markets; but at all times they want a round price for them, if they be not old and worn out. In that, of course, you exercise your judgment.

Still there are many gentlemen who take great delight in breeding Spanish fowls for fancy; and others again who take equal delight in breeding Dorkings. Such gentlemen spare no expense in procuring the finest specimens that either friendship or money can command. I have heard of thirty guineas being given for a brace of Dorking pullets; and also a higher price for a celebrated Spanish cock, because he was up to the full mark of the fancy.

These gentlemen, in the course of breeding, have many birds, both cocks and hens, that do not come up to the standard of the fancy, and are consequently cleared out from the stock. Now these are just the very birds for you, and will suit your purpose as well as the finest birds in the roost. It is only breed and size you want; and I am satisfied that these gentlemen, knowing the purpose for which such birds are wanted, will feel a patriotic delight in benefiting their native land by placing these outcast chickens within reach, and thereby, not only increasing the prosperity of British farmers, but promoting national independence, and augmenting the domestic resources of the country.

This subject merits the most serious consideration of the landed gentry of England; for, by furnishing facilities for their tenantry in obtaining such a valuable acquisition to their farm stock, they will not only assist them to clear their way in ungenial and less favourable seasons, but contribute much to that good feeling that should always exist between landlord and tenant, and thus ensure the happiness and domestic comforts of farmsteads in general.

Before entering into my intended calculations, as to the profits arising from Spanish Dorkings, I shall take the opportunity of noticing an extraordinarily prolific hen, belonging to a gentleman in Somersetshire. In 1852 she was seven years old; and in six years laid the following number of eggs:—In the first year, 142; in the second, 200; in the third, 160; in the fourth, 144; in the fifth, 134; and in the sixth, 147; thus making, altogether, 927 eggs from this single fowl. But this is perhaps an extraordinary case, and cannot be any certain criterion as a rule.

Let us now consider which is the most profitable to farmers, keeping rats or Spanish Dorkings.

In the first place, we shall set down a rat to eat and waste only a wine-glassful, or half a gill, of wheat in twenty-four hours; and let us suppose that wheat is at the rate of sixty shillings per quarter. That would be a pint among eight, or a quart among sixteen rats per day! Now that quart of wheat, at the rate of sixty shillings per quarter, would cost over twopence threefarthings.

In the second place, let us remember, that you do not feed your fowls with corn at sixty shillings per quarter, but with the refuse from thrashing. However, let us suppose you feed them with good sound barley at forty shillings per quarter; then, for twopence three farthings you could give sixteen fowls about three pints of good barley per day. Now if sixteen fowls had daily three pints of good barley, besides what they would pick up about the farm, do you not think they would do well? In a word, do your fowls get half that? If they do, they get more food than farm fowls in general.

In the third place, 125 rats, allowing only three glasses for waste, eat a peck of corn per day, and thereby cost you at the rate of £34 4*s.* 4½*d.* per year. Bear in mind also, that they yield you nothing in return, except a host of their young to keep. So we find that, by keeping 125 rats, your £34 4*s.* 4½*d.* is infinitely worse than if thrown behind the fire, because in that case you would know the end of the loss; but as it is, you can form no idea whatever as to the end of the losses by rats, except by their extirpation.

Now let us suppose you have destroyed all the rats, and have 125 Spanish Dorkings in their place; then let us see what the difference would be. In the first place, let us set them down to lay only 120 eggs each in the year. That will be at the rate of four eggs per week, for seven calendar months: thereby leaving five calendar months, or twenty-two weeks for sitting, moulting, and so on. Now this is fourteen eggs less than the hen above mentioned laid in her worst year, and eighty eggs less than her best year. However, to keep the matter within the mark, let us set them down to lay 120 eggs each in the year. In the second place, what shall we sell these Spanish Dorking eggs at? Here I must tell you that a French-egg merchant of London, after weighing Aunt Jane's eggs against his own, bought a number to sit under his fowls, to raise eggs for his own eating. This was in the middle of May, and he was then selling his largest foreign eggs at twelve a shilling; but if you picked them, he would not let you have more than ten for the shilling. He sold them at all prices, according to the size; and as to the small stale ones, you could have them from twenty to twenty-four for the shilling; but they were only the size of bantams' eggs; nor would he change them if rotten.

French and English Eggs.

Here, I must ask you this simple question—If steam can bring foreign eggs and poultry from abroad to London, and then from London convey them by railroad to nearly every town in the provinces, pray cannot steam convey British eggs and poultry from all parts of the country to the London markets? Yes: and what is more, you have this advantage over the foreigners,—all your eggs can be new-laid; for no matter in what part of the country they may be, your fowls may lay their eggs this morning, and to-morrow morning the Lord Mayor of London may eat them! Or the Prime Minister of England and his political friends may regale themselves with British new-laid eggs, and such as the wide world cannot surpass for size and quality! What is still more, I am happy to believe that there are but few English ladies or gentlemen who will allow either foreign eggs or poultry to come to their tables, if they know it. Nor would the middle or industrial classes partake of them, if they could obtain English produce as cheap. Here, then, the great question rests, that you can sell an infinitely superior article as cheap, and a great deal cheaper than the foreigners, if you breed and feed your poultry as did Aunt Jane. Beyond what I have told you there is no secret in the matter; and what Aunt Jane did you can do, if you have the same determination and perseverance. The end of that determination and perseverance will be, that the foreign produce will find its way into their own markets, to the benefit of their own population, while the millions of British money which now leave our shores to enrich a few speculative individuals, will come into the hands and pockets of British farmers and their families.

In speaking of London, I merely speak of it as being the great emporium of consumption. Still, what applies to the metropolis, in this instance, may equally apply to other places and ports in England, according to their size, station, and commerce. But are you aware, my friends, that in despite of all the foreign eggs imported, the hotel keepers, buttermen, and greengrocers of London often contract with fowl keepers to give them twopence, and twopence halfpenny each, the year round, for new-laid eggs, and then make a handsome profit out of them? As for hotel keepers, many of them enter new-laid eggs in their customers' bills at sixpence each. However, I shall not talk of eggs at sixpence each, but of French eggs at sixpence a pound.

Here is a matter to which the attention of farmers and the public in general has never been properly directed; and that is, the consideration of the weight of eggs, as well as the number. Many persons innocently believe, when they see a large number of French eggs for a shilling, that the eggs are very cheap; but that is not a just criterion, as the following will clearly prove. When an egg-merchant sorts out a chest of French eggs, he knows by practice which basket an egg belongs to, the instant he takes it in hand. Each egg is sorted out according to its size, and when sorted it matters not which basket you choose from; for if you weigh a shilling's worth out of each, separately, you will find each shilling's worth, as near as may be, the same weight; that is to say, that those at twelve, fourteen, and sixteen a shilling, are as heavy as those at eighteen or twenty a shilling. Of course eggs are not like cheese or butter, where you can cut off an ounce or so, or stick half an ounce on, to make them exact weight. But take French eggs on the average, and you will find them all selling at a trifle over sixpence a pound, or two pounds for a shilling. I have frequently weighed them myself, and have also had them weighed at various times in various shops; but I have always found them varying from sixpence to sixpence halfpenny a pound. This is a matter any one can prove for himself.

Thus we find that stale foreign eggs are sold at sixpence a pound, or two pounds for a shilling. Now what can British farmers sell their new-laid eggs at? Can they sell them at sixpence per pound? But bear in mind that this is the price in the full laying season, not in the fall of the year, or in the winter. At those periods, foreign eggs fetch somewhere about ninepence per pound; and as for English new-laid ones, they will fetch their weight in copper. Still, to calculate with certainty, let us not reckon these extra profits, but lump all your eggs together at one price; and let that price be the lowest that foreign eggs are retailed at, namely, sixpence per pound. Let us also divide your eggs into two sizes, namely, six to the pound, and eight to the pound. That would average seven to the pound. But to make the matter easily understood, let us set them down at twelve and sixteen for the shilling.

Now can you sell your new-laid eggs at sixpence per pound, or two pounds for a shilling? Perhaps you have never given the matter a calculation; therefore I will calculate it for you.

In the first place, let us calculate the cost and profit of one hen; then we can multiply by any number we please.

In the second place, we will set down a hen to cost no more keeping than a rat; that is, supposing a rat to eat and waste no more than a wine-glassful, or two ounces of corn per day, and valuing the corn at sixty shillings per quarter. Then that cost would allow the hen to have a glassful and a half, or three ounces of barley per day, at forty shillings per quarter. But this of course is quite independent of what she might forage out and pick up for herself.

In the third place, we cannot pretend to calculate to a fraction; still a wine-glassful and a half per day for twelve months would amount to one bushel, four pints, and three glassfuls. But considering the amount and cheapness of the soft food necessary during the moult and winter months, it would bring the cost considerably under the bushel. However, let us set it down at a bushel of barley per year; and then, barley being at the rate of forty shillings per quarter, it would leave five shillings for the bushel.

In the fourth place, let us say nothing about the chickens the hen would have; but set her down to lay at the rate of four eggs per week for seven calendar months, or thirty weeks out of the fifty-two; and they would amount to 120. That would be exactly one halfpenny each; or, averaging them at seven to the pound, they would weigh seventeen pounds, and one egg over; thus costing you threepence halfpenny per pound, or a shilling for four-and-twenty eggs.

Thus I have been calculating that you buy their food at the market, and pay at the rate of forty shillings per quarter for it. But you, gentlemen, growing your own barley, of course have it at prime cost. Now does barley cost you forty shillings per quarter, growing? Or rather, when rent, tithes, taxes, tillage, seed, and wages are paid, does it cost you twenty shillings? I think not; but let us set it down at twenty shillings per quarter! and then, gentlemen, pray what becomes of the tail-ends after winnowing the corn? that, I believe, always falls to the fowls' share. However, to make the matter plain, let us take barley, tail-ends, meal, potatoes, sharps, and butcher's offal, and lump them all together at twenty shillings per quarter, and then your eggs will stand you at one farthing each, or a penny threefarthings per pound; that is, at the rate of about three pounds and a half for sixpence, or eight-and-forty eggs for a shilling.

All these calculations tend to demonstrate the cost and profit of a hen to those farmers who have no rats. But to the farmer who has always been in the habit of keeping a rat, and who is determined to do away with it, and keep a good hen in its place, the hen is infinitely more than all profit. I say more than all profit, because she not only lives well upon what the rat is accustomed to waste, but clears the farm of grubs, worms, slugs, insects, &c. Besides that, she does not give him a family of hungry young rats to keep.

Still, to make the matter of cost and profit more clear, let us set down her food at five shillings. Then set down the eggs at sixpence per pound; and who would not sooner give sixpence a pound for English new-laid eggs, than the same price for foreign stale ones? We next calculate the hen to lay 120 eggs, which, at seven to the pound, weighs seventeen pounds, and one egg over. Now seventeen pounds of eggs, at sixpence per pound, amounts to 8*s.* 6*d.*; and if we value the hen herself at only two shillings, then the hen and eggs produce 10s. 6d. Now let us deduct five shillings for the bushel of barley, &c.; and then does it leave just 5*s.* 6*d.*, or 110 per cent. profit. But to farmers who grow their own food, and who of course have it at prime cost, the expense of a hen will be only 2*s.* 6*d.*, which will leave 8*s.*, or 320 per cent. profit. But to those farmers who have bestirred themselves, and now keep a host of fowls instead of rats, they are all profit together.

Thus the difference between keeping 125 rats, and 125 Spanish Dorkings, will be, that according to the wine-glass standard the rats will cost him at the rate of £34 4*s.* 4½*d.* per year; whereas the fowls, in the same period, will bring into his pocket £50.

And now let me show you the difference it would make to each, if you did away with your rats, and kept Spanish Dorkings in their stead.

In the first place, let him who estimates his rats at 250, do away with them, and keep the same number of Spanish Dorking hens; then, instead of loss, he will realize over £100 clear profit in the year.

In the second place, let him who sets down his rats at 500, bestir himself, and keep 500 of these hens instead, and he will find himself a yearly gainer of about £200 clear money.

In the third place, let those who estimate their rats at 750 in number, follow the same example, and they will find the yearly profits of their hen-roost to exceed £300.

And lastly, to those who estimate their rats at a thousand, the difference will be, that by the rats they lose a vast amount of money, and by the fowls they will clear about £400 yearly.

Embryo Eggs.

In concluding my remarks on the breeding, feeding, and value of poultry (all of which are the result of practical experience), allow me to say a few words to the querulous. I have been reminded that rats do not increase by figures, but by the amount of food afforded them; and also that fowls do not lay or increase by arithmetical rule. I can assure my friends, that I not only clearly appreciate the wisdom of this opinion, but assure them that I am perfectly aware that neither herbivorous, granivorous, or carnivorous animals can either increase or live without food; also that in a domesticated state their increase is in a great measure regulated by the amount of seasonable nutrition that we afford them. Still, for rats, there is this to be said, they do not either ask our leave, or wait our permission, but help themselves; and when our supply runs short they pay their addresses to some one else. Thus are they always well supplied with the best of food, and, in the usual course of nature, increase and multiply at pleasure. But as to fowls, they cannot gnaw or work their way into ricks, barns, and granaries, like rats; consequently they are mainly dependant upon us for assistance. I am also satisfied that an egg is not the result of chance or accident; but that a certain number of embryo eggs are formed in the pullet before the pullet is hatched, or rather with her first formation in the shell; and that no after circumstances whatever can form new embryo eggs. That the maturity of these infantine eggs is regulated entirely by the amount of nutrition regularly afforded the pullet for its gradual growth and completion. Hence it is that myriads of fowls live and die without laying half the complement of eggs that nature has implanted within them, because they have not had food enough to bring them to perfection; thus proving thereby, that the great economy of fowl-keeping consists in feeding them well and regularly, and then a hen may lay her full complement of eggs in about three seasons; but if irregularly and indifferently fed, she cannot possibly complete her allotted task. Here, then, is the real cause of complaint with many who keep fowls, and declare them unprofitable. They give them just enough of food to keep them alive, and to keep them from laying; and thereby the food is of course all wasted. Now if this be not a penny-wise and pound-foolish philosophy, it is at least what the good people of Yorkshire would call, "Spoiling a pig for a ha'poth of tar."

The Incubator, or Egg-hatching Machine.

In my humble opinion this machine is a most valuable invention, both for farmers and poultry breeders in general, because there is no chance of addled eggs; since every perfect egg will surely come to maturity, and there is warm and dry accommodation for the young chickens till they are fit to be taken away. Neither are there any chances of the eggs being broken or spoiled from the fractious disposition of the hen, or by her deserting them, or quarrelling and fighting. Indeed the machine surpasses hen-hatching altogether, both for the quantity and quality of the chickens it produces. Then, again, there is this advantage, you will not want your hens to sit; consequently, when they begin to cluck, just douse them in a bucket of water, and in two or three weeks they will come on again to lay; thereby saving both time and money.

Still, these are matters I merely lay down for your consideration. But, under any and every consideration, get rid of your vermin, or I can hold out no hopes whatever, either of pleasure or profit from your poultry-yard, even though you might purchase the most noble specimens of birds the world ever saw.

The superior Profit of Poultry over Beef, Mutton, and Pork.

Now allow me to ask you this simple question—Pray, can you feed beef, mutton, or pork at a penny three farthings per pound? Of course you cannot; and why? Because in feeding oxen, sheep, and pigs you have only their carcasses to pay for all! while in poultry-keeping you not only have the hen's carcass, but all the eggs she lays in addition; which eggs, in three years, would produce more than twelve times the value of the hen herself! that is to say, were the hen herself to fetch two shillings, her three years' eggs would fetch £1 5*s.* 6*d.*, showing the vast superiority of poultry-keeping over every other consideration in live stock for farmers.

But bear in mind, gentlemen, that in all my calculations I have not said one word about the superior price that eggs fetch in the scarce seasons of the year, or of the chickens and capons you would sell for the table. I have put down everything at the lowest calculation, in order to set at rest all querulous argument. Surely you must see how you have the battle in your own hands, and can beat the foreigners back into their own markets, not only to your benefit, but also to the benefit of their own population. You have only to force the contest, and conquest is certain; for should they sell their eggs at fourpence halfpenny per pound, you could sell yours at threepence halfpenny; and then make cent. per cent. or one half profit, even though you had to pay twenty shillings per quarter for the food! But, as it is, to you they would be all profit together; consequently, if foreigners brought their eggs down to a penny per pound, you could bring yours down to a halfpenny, and then be that halfpenny the gainer.

Thus you see that victory is certain, if you will only buckle on your armour and set to work like Englishmen in earnest.

Extraordinary Eggs.

"Uncle James, or rather Mr. James Rodwell, author of 'The Rat, and its Cost to the Nation' (says Bell's

Weekly Journal), "called at our office, and presented us with two eggs laid the same morning. The first egg, which was of the ordinary size for his fowls, measured six inches and one-eighth round the oval, and five inches and one-eighth round the girt, and weighed two ounces and three-quarters, thus averaging six eggs to the pound. But the monster egg measured seven inches and seven-eighths round the oval, six inches and one-eighth round the girt, and weighed over three ounces and three-quarters. Last season he had three of these monster eggs, with single yolks, the largest of which measured over eight inches round the oval, six inches and a half round the girt, and weighed over a quarter of a pound. His fowls are bred between the Spanish and Dorking, and have supplied him and his family with new-laid eggs for breakfast since the beginning of December last. He recommends all who are keeping fowls for profit, to take advantage of the present season, if possible, and breed their chickens between the Spanish and Dorking, as they are excellent layers, and a fine fowl for the table. Still, he says, it must be borne in mind that the height of economy in fowl-keeping is to see them daily well fed; for it is a fact beyond all dispute, that the more substance they consume, the more regularly will they lay."

Pigeons and Farm Birds in general.

As to the various birds that feed upon the farm (with the exception of pigeons of all kinds), I have not one word to say against them, but quite the contrary; for I believe, that where they rob the farmer of one grain they repay him infinitely more than a hundred fold, by devouring myriads of grubs, caterpillars, and insects that would otherwise eat out the hearts both of grains and plants. But as to the pigeon tribe, with the exception of the little service they may render in the autumn by helping to clear the ground of the scattered grain after the harvest, I believe them to be most destructive creatures, in carrying away turnip seeds, &c., in their crops by hundreds and thousands at a time; and they never touch either snail, slug, worm, grub, or insect. Whereas, on the contrary, fowls will not only devour all they can catch, but will hunt out every hole and corner to find them; and what, perhaps, is not generally known, is the fact that they will kill and devour both mice and young rats. Nor are they very backward in having a taste of a full-grown rat, if it lies dead in their way. I have seen a perfect scramble for a mouse or young rat that was dodging in and out among their legs; and after one had pecked it, and a second, a third seized it in her mouth, or bill, and ran away, when there was a regular race for its carcass, which among them was torn to pieces and devoured. As I have told you before, no kind of animal food comes amiss to them; for they will not only eat the flesh of any animal ranging between a horse and a mouse, but will pick the bones of their own species, or any other bird, with the utmost relish and satisfaction.

As to pigeons, however, they live on nothing but grain, and will fly for miles in quest of it. Indeed no grounds are safe from their depredations. I read of one case wherein curiosity led a gentleman to open the crop of one he had just shot, and he counted over four thousand turnip seeds it had swallowed. In 1853, a wood-pigeon was shot in a field of wheat at Newhouse, in the parish of Dalry, N.B., when, on examination, the crop was found to contain 858 grains of wheat. Surely farmers should be on the alert to put an end to these destructive creatures.

And now I have only to express a hope, in conclusion, that I have acquitted myself to your satisfaction, in showing the devastating character of the rat, and its cost to the nation. It, therefore, only remains for you to do your duty, and carry my plans into execution; and by so doing, you will not only enrich yourselves and your families, but do much towards increasing and sustaining native independence, and the honour and dignity of our native land.

THE END.

COX AND WYMAN, PRINTERS, GREAT QUEEN STREET, LONDON.

Printed in the United Kingdom
by Lightning Source UK Ltd.
105825UKS00001BA/19-21